For John Graham, whose help and guidance enabled me to start this long and wonderful journey with gardens.

NEW EXPANDED EDITION

PAUL BANGAY'S GUIDE TO PLANTS

Photography by Simon Griffiths

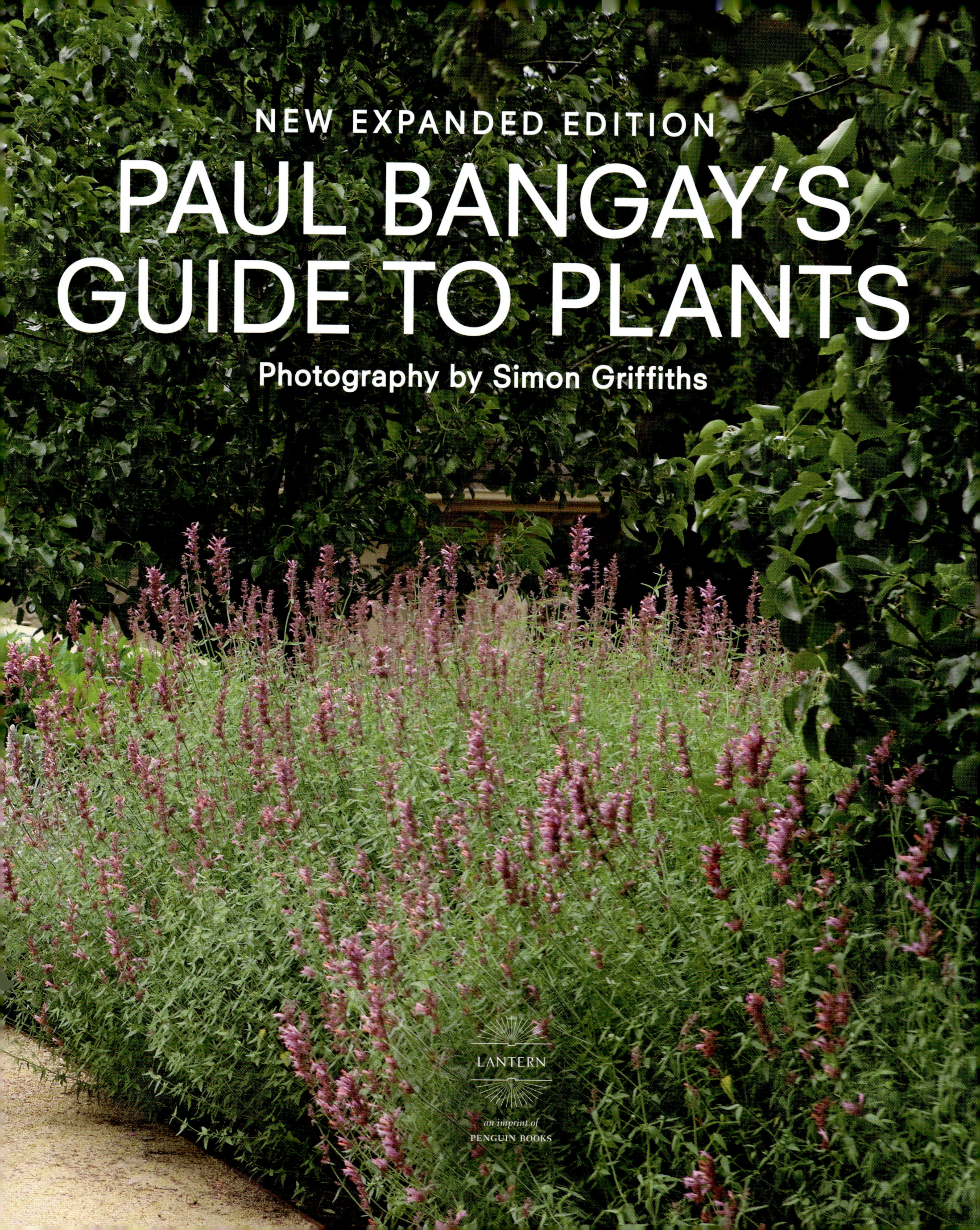

LANTERN

an imprint of
PENGUIN BOOKS

Contents

Foreword

It has been 10 years since I researched and wrote the first edition of this book. In this time, much has changed in both gardening styles and climatic outlook. The world now considers climate change as a real and rapidly approaching threat, and our plant choices must change so our gardens can adapt to this drier and hotter world.

I have included a chapter on succulents in this new edition for this reason. Previously, succulents were never on my mind when designing planting schemes. I am now increasingly using them in my designs – not all types of succulents, but especially the choices included in that chapter. This is merely one way in which I am adapting to climate change. Another strategy is to take a very critical review of plant selection when designing. For many in Australia, more delicate and cool climate plants such as hydrangeas and azaleas (the much-loved plants of our parents and grandparents) are now no longer suitable. Before selecting your plants, ensure you carefully consider the water requirements and optimal climate for each plant description in this book.

Landscaping styles may evolve on a much slower cycle than interior styles, but evolve they do. Over the past 10 years, there has been a noticeable shift towards a more relaxed and casual approach to gardens. 'Wild' and 'natural' are words used to describe many contemporary gardens now. I personally have been adapting and softening my designs for many years now. Gone are the rigid box-bordered geometric garden beds in favour of more organic and relaxed shapes, with groupings of spheres or organic plant forms in accent areas of the garden bed. The new chapter on clouding illustrates all the plant types that can be used to create these forms that have now become hugely popular. The exciting part of this chapter is that many of these plants are Australian natives, such as *Correa* and *Westringia*. Integrating more native plants into my designs helps with climate change adaptions as well as merging cultivated gardens with the surrounding natural landscapes.

Of course, with the exception of a few plants that haven't adapted to a changing environment or to gardening fashions, all my favourite and well-used plants are still included in this edition.

Opposite: Herbaceous perennial garden in the pool garden at Stonefields

Introduction

Over the years I have worked in the far reaches of Australia, designing and creating gardens for compact inner-city courtyards through to sprawling rural estates. Each project throws up fresh challenges, requiring a careful balance of proportion and scale within the confines of the landscape available. But by far the most important decision of the whole process is choosing which plants to use and where.

For example, I love to plant pleached hedges of ornamental pear (*Pyrus*) or linden (*Tilia*), but neither of these will work in Brisbane. And I've always been fond of formal clipped English box (*Buxus sempervirens*), but it would be absurd to incorporate it into a tropical garden.

So how do I select the right plants for my designs?

When I first visit a site for a new garden, I always check what grows well in the local botanical gardens, and I can often be found peering into neighbours' gardens to examine their plants, especially those that are doing well. It also pays to take note of the vagaries of the climate at your particular site. Sometimes a plant might do well in a neighbour's garden but wither and die in your own because of a particular microclimate that occurs in that position. The specific climatic conditions that you experience at your place, combined with how much water is available, are the two most important factors when deciding what to plant.

Once I have observed the climate and microclimates of a site, that's when I start to draw on my list of reliable favourite plants which I have compiled over the years and which is contained within the covers of this book. There have been countless additions as I've discovered new plants that love Australian conditions, and I've also taken many plants off the list that just can't cope with our extremes of temperature. All the plants included here have been used successfully in gardens I have designed throughout the country.

This is a very personal collection of favourite plants. It is not meant to be an exhaustive plant reference guide – there are plenty of those. Rather, it is a sort of catalogue of suggested plants for those who like my gardens and would like to create something special in their own plot.

Opposite: *Ligustrum vulgare* 'Tuscan', *Carpinus betulus* 'Fastigiata'

I have to acknowledge that the plants I most enjoy working with are those that enjoy cold to warm temperate conditions – plants that do well in the south of the continent and north up to Sydney and rural New South Wales. I adore all the wonderful perennials and bulbs that require a cold winter and no humidity in the summer, and for this reason you'll find more plants in this book that fit these criteria than ones which thrive under tropical or subtropical conditions. Apologies to my friends up north, but I hope you will find enough here to inspire you.

My ultimate goal is to create gardens that stand the test of time, eliminating the need to replant every 10 to 15 years. As with the architecture of the house, the garden should remain. This is how I would like to see my gardens, remaining in harmony with the buildings, mellowing with time and maturing into a place of timeless elegance.

Of course gardens do evolve with time and one part of this process is that the minor plants change; this may be due to increased shade as trees mature or simply because you want to experiment with other plants or colours.

Do not be afraid to pull out plants that are not doing well. A garden is not a hospital for sick plants. The garden should give you the joy of seeing a thriving green space, where nature meets you at home. If a plant is not working in the space it's been given, relocate it or discard it altogether.

This book is very much a companion to my previous title, *Paul Bangay's Garden Design Handbook*, in which I touched only lightly on plant selection. While the design book takes you through the methodology I use in creating my garden spaces, this book gets down to the nitty-gritty of why we garden in the first place – the plants! My hope is that this collection of my very favourite plants, gloriously photographed by the wonderful Simon Griffiths, gives you plenty of inspiration to experiment with plants in your own garden.

Opposite: *Cupressus macrocarpa*, *Rosa rugosa* var. *alba*, *Buxus sempervirens*

Before you plant

Planting schemes

In this book we really are beginning at the end. The planting scheme is the final piece of the puzzle in creating a garden, not the first. For me, plants complement and enhance the overall design of a garden but they do not drive it. I want the architectural elements in my garden designs to last a very long time, while the planting scheme may chop and change and even be lost with the passing of time.

Having said that, when I first look at a site it often becomes clear to me very early on how the garden should look, including the overall feel of the planting scheme. I rarely lock in the selection of individual plants until the garden structure is laid out, but there is something about each site – 'the genius of the place', as English poet Alexander Pope put it, and as great landscape designer Russell Page was fond of quoting – that informs me of how the garden should look and what it needs in order to thrive. Is the garden contained? Is it influenced by views to other planting schemes or neighbouring properties? What is its aspect? What is the architectural style of the house? These are the starting points for choosing a successful planting scheme.

Of course the reality we must have at the forefront of our minds when considering plant selection is the climatic conditions of the particular site. In most cases irrigation is just not sustainable, so choosing plants that work with the natural rainfall of the location is often paramount.

Another limiting factor when creating gardens in subtropical to tropical areas of the country is the amount of humidity. Often clients in areas of moderate to high humidity desire soft, flowery planting schemes more suited to the southern states. Roses, irises and lavender just will not thrive in these situations and it's best to acknowledge this at the start to avoid disappointment. If you live in an area of high humidity, take a careful look at the climate information provided for the plants in this book and in others before investing time, effort and money in plants that require drier conditions.

For many people the garden is not their priority and they require low-maintenance planting that will still offer them a satisfying outdoor green space. This is when evergreen perennial plants and shrubs are the order of the day.

For the rest of us, a planting scheme might be a very personal selection of the plants that give us the most joy. People who are obsessed with their gardens tend to enjoy collecting rare plants and will have very particular tastes about individual plants. For readers who fall into this category, there might be some plants in this book that they detest or there may be plants they love that are not represented here. We all have our personal tastes. The best planting scheme is one that reflects the personality of the owner and I hope this book will be used as inspiration rather than as a rulebook.

When considering a new planting scheme for your garden, think carefully about whether you require flowers and scent. Do you want a simple, formal green garden or are you after a more casual garden full of flowers? Flowering gardens are always more difficult to design for inner-city spaces as flowering plants usually do not look great when not in bloom and it's nigh on impossible to achieve a flowering garden all year round. But of course there is the joy of finding flowering treasures in your garden that please the senses and are all the more valuable for their temporary nature. A formal evergreen garden is often an excellent option for urban locations. Totally green gardens are easy to design and easy to maintain. Even a garden consisting of multiple layers of hedge requires only a clip once or maybe twice a year and little more in the way of maintenance.

The big question is, has a uniquely Australian planting scheme evolved yet? We know what a French scheme looks like (where would it be without the lavender?) and many in this country have incorporated Italian planting schemes in their garden (think English box, wisteria and pencil pines). I do believe that by incorporating a number of native Australian plants into our designs we are slowly developing our own planting scheme. Our casual way of living also contributes to a less formal approach to our garden designs, and as our modern Australian architecture matures in style so will our landscape styles. For me I love using our native brush cherry (see page 68), lilly pillies (see page 21) and figs (see pages 37 and 41) as hedges. Tussock grass (see page 249) and native rosemary (see page 144) I use in the beds, while happy wanderer (see page 183) is a favourite of mine for a groundcover or climber. Incorporating these native plants into gardens which also contain exotics that suit our conditions provides me with a planting scheme I consider to be modern Australian.

Opposite: *Platanus orientalis*, *Rosa rugosa* var. *alba*, *Buxus sempervirens*
Below: An example of a planting scheme sketched out for a client of Paul Bangay's

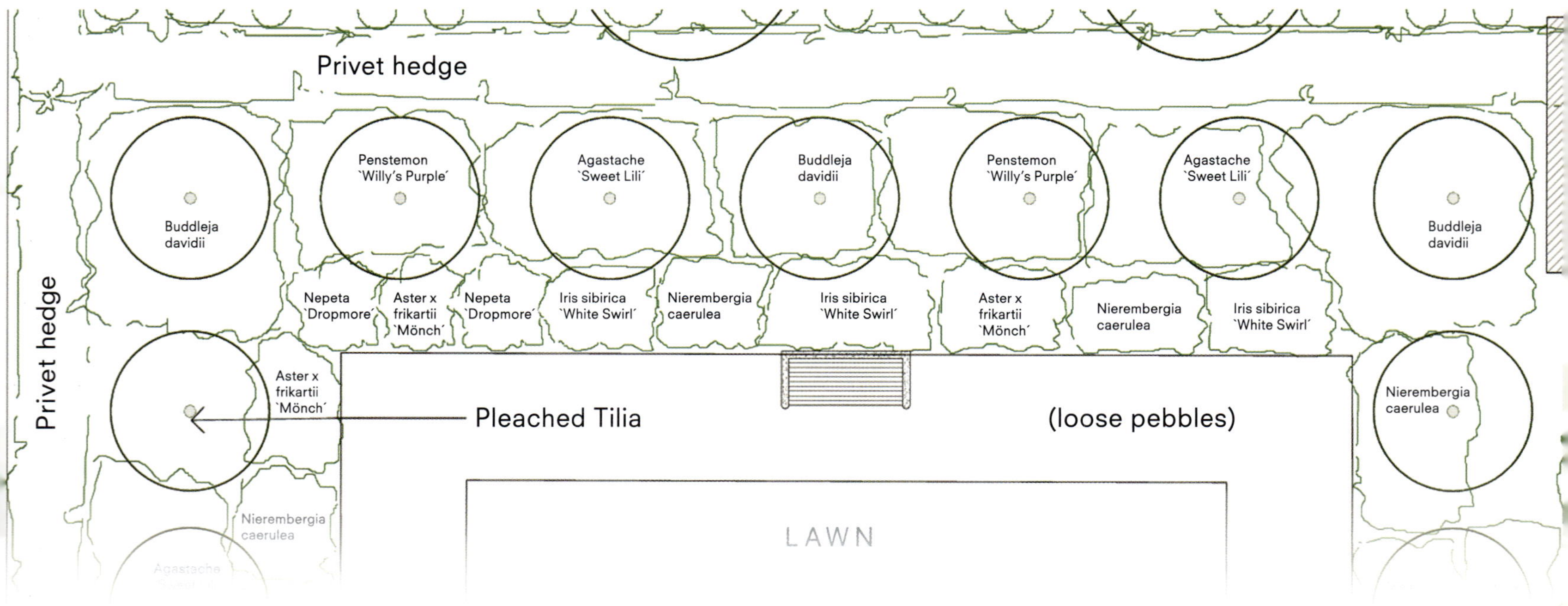

Colour schemes for the garden

When I first started creating gardens, 25 years or so ago, everyone wanted a white garden. This was perhaps in imitation of Vita Sackville-West's magnificent gardens at Sissinghurst Castle in Kent, a colour scheme revived by the Australian gardener Lady Joan Law-Smith at Bolobek, which sits at the base of Victoria's Mount Macedon. Today, we are all more willing to experiment with colour in our gardens – even with the once much-despised red. In one part of my country garden I use a colour scheme of deep magenta and rich red wine hues. It's liberating to play with a diversity of colour after years of restraint.

It's true to say, however, that choosing successful colour schemes for plantings in the garden is not always straightforward. While there are some useful reference points you can use when considering a colour scheme, there are no hard-and-fast rules to follow. It might come as second nature for experienced gardeners and landscape designers, but choice of colour schemes is much more of an art than a science and sometimes things that look good in your mind's eye just don't seem to work in reality. In many cases it comes down to experimentation, more experimentation and still more experimentation.

Starting points to get you thinking about a colour scheme for your garden might be to consider the colour and architectural style of your house. White houses often look good with predominantly white and green gardens. Modern houses frequently suit a splash of warm colour such as red. Think about the colours used in the paving – greys and black allow for deep, dramatic colour schemes including purples and deep reds.

Occasionally I have successfully created colour schemes using furnishing fabrics from the inside of the house as the inspiration. In one such scheme, the curtains that framed the view through to the garden outside were a deep pink and purple chintz. So I created a garden just outside the window that contained deep pink roses and purple irises. The effect from inside the house, from where the garden was most often viewed, was one of elegant flow and unity.

It's worth considering also the effect that light has on the space. A dark area, such as a south-facing courtyard, can be brought to life by a light colour scheme.

Left: *Veronicastrum virginicum* 'Diane' Centre: *Leucanthemum* 'H.M. Allsop' Right: *Ammi visnaga*

Vita Sackville-West's famous 'White Garden' was intended to be viewed by moonlight, when the white flowers were iridescent against the dark of night.

If your site is subject to intense light, then deep, rich colours can often work well. Greys and purples seem to suit the harsh Australian light and many Australian native plants display these tones. However, white should not be completely disregarded as it can offer a psychological 'cooling' effect.

Most clients of mine have played it safe with the colour schemes they chose, preferring whites, greens, blues and pale pinks. But in a few noteworthy projects I've been able to reflect the personality and mood of the client in a bolder colour scheme, which I must say has been really satisfying for both them and me. One such was a client on Sydney Harbour who was mad for the colour pink. The interior of her house was pink, the exterior was pink, she even wore pink clothing. Of course, the planting colour scheme needed to include pink too. We opted for all pink and white. The colour scheme worked because it reflected the personality of the owner and it became a deeply personal garden. I loved it for this reason.

For me, the thing I value most in my garden is a sense of peace and tranquillity. For that reason I constantly come back to the most relaxing, tranquil and at the same time vibrant colour scheme with an all-green garden. This scheme relies on the subtle variations in the colour according to the time of year. Nothing beats the electrifying green of new growth on the English box plant (see page 27); from its dull, safe, deep green bursts a flush of apple green that brings the whole garden to life.

If you live in tropical or subtropical areas of the country, however, the most successful plants for the region tend to be big, bold, bright colours. Bougainvillea (see page 177), clivia (see page 358), strelitzias (see page 367) – there's no point shying away from them when the climate dictates a bright colour scheme.

So how do you choose your own colour scheme? After considering the style and colour of the house, the aspect and climate of the garden, bring it right back to the personal: what do you enjoy? What mood do you want to convey in the space? Make it your own, avoid a clutter of colours, and your garden will sing.

Below: *Buxus sempervirens, Nepeta* 'Dropmore', *Agastache* 'Blue Fortune', *Agastache* 'Sweet Lili', *Aster* x *frikartii* 'Mönch'

Selecting the right plant for your site

Here are the rules I stick to when I'm choosing a plant for a particular location in the garden.

1. Know your climatic conditions

You can do this by checking which climate zone you are in for maximum and minimum temperatures as well as for rainfall. We have included a simplified version of climate zones on page 14. For more detailed info, look up 'climate zones' on the Bureau of Meteorology website (bom.gov.au). Check the suitability of the plant you are interested in against the climate zone/s you are in.

You can also learn a lot by simply observing what is growing in your area, in neighbours' gardens and in local botanical gardens, or by visiting your local nursery and talking with the staff there about what works well. Nothing can replace local knowledge.

2. Consider the position

How shady is the position you are planting in? As much as you might love the idea of including feather reed grass (see page 222) in a particular spot in the garden, if the plant won't tolerate shade but the position is overshadowed by a house, fence or tree, you're better off choosing a shade-loving plant such as hellebore (see pages 230–31). Planting a sun-loving plant on the south side of a structure in the shade will lead to poor or no flowering or leaf growth. Just not worth it.

Another important consideration is the amount of moisture available in the soil. Most plants like a well-drained position. If the position you are planting in is always damp, it's best to recognise it at the outset and select a plant such as bog sage (see page 255), which will thrive under such conditions.

To determine how much water is held in your soil, dig a small hole. In very boggy soils, water will seep into the hole and fill it up. If you fill the hole with water and it takes a long time to seep away, again, it's a soil with poor drainage. But if the water seeps away quickly, you've got excellent drainage.

3. Check the mature height and spread

No doubt you know a story or two about a neighbour with a 15-metre jacaranda in their courtyard or a plane tree planted directly adjacent to a house and the disastrous consequences this can have. It's wise to be realistic about just how big the tree or plant will get before you plant. Obvious, but a trap that so many fall into. Enough said.

4. Select a tree according to its deciduous or evergreen nature

Consider carefully the purpose of the tree you are planting. I like to plant deciduous trees near a house to provide winter sun and summer shade. Deciduous trees also provide the garden with much-needed colour in the autumn. However, evergreen trees do provide interest during the gloomy winter months. I find I tend to plant more evergreen than deciduous trees in my schemes but I would have to say that getting the right balance is the key.

5. Consider drought tolerance and water availability

Drought tolerance is a major consideration in Australia and it's not going to go away. Along with occasional extreme floods, climate scientists tell us we can expect periods of drought. We need to be careful to include more and more drought-tolerant plants and trees in our gardens. You might be surprised by the hardiness of some of the plants I've chosen for this book – the oak trees, for instance, which are extremely hardy.

If you do decide to include some moisture-loving plants in your scheme, be sure that you have an ample water source such as a very generous-sized rainwater tank and/or a greywater system.

6. Make your nursery work for you

Time and time again when travelling to a different location to design and install a garden I have found the knowledge of local nursery staff invaluable. They are always very in tune with local conditions and usually have a vast database of plants that will grow well in that area.

One word of advice: challenge your local nursery. You don't need to settle for the hybrids that are pumped out by the wholesale nurseries. The more you read in books such as this and in magazines, the more you are able to ask your nursery to source different and more interesting plants for you.

Climate zones

Our continent is vast and there are so very many complex climate zones within it that it is very difficult to write a book about plants that suits everyone everywhere. I have tried to simplify this by dividing the country into zones and referencing this to each plant description.

Keep in mind, though, that this is a rough guide only. It does not take into consideration climate variations that can occur particularly in mountain regions or even within a single large garden space with artificial irrigation. I am often surprised to see plants flourishing in regions of the country where technically they should not grow.

Above: *Allium giganteum*
Opposite clockwise from top left: *Viburnum opulus*, *Gardenia augusta* 'Florida', *Plumeria acutifolia*, *Bougainvillea*

Key to climate zones

Cold

Mild to warm summers where fewer than seven days per year reach 30°C. Heavy frosts and snow in winter. Generally uniform rainfall throughout the year. Includes most of Tasmania, high areas of the Southern Tablelands, the Australian Capital Territory, and high areas of the New South Wales Northern Tablelands such as Armidale.

Cool temperate

Warm summers where the average January maximum temperature is less than 30°C. Frost occurs in winter and can continue into spring. Suitable for growing plants that require a defined period of chilling. Wet winter and low summer rainfall. Includes most of Victoria (except Melbourne), the Albany region in Western Australia, and the south coast of New South Wales.

Warm temperate

Warm summers where the average January maximum temperature is less than 30°C. Light to no frost in winter. Not usually suitable for growing plants that require a defined period of chilling. Reliable rainfall throughout the year. Includes Sydney, Adelaide, Melbourne, Perth.

Subtropical

Warm, humid, wet summers where the average January maximum temperature is less than 30°C. Mild winters with no frost. Not suitable for growing plants that require a defined period of chilling. Low winter rainfall and reliable summer rainfall. Includes eastern seaboard, from north of Brisbane south down to Coffs Harbour.

Tropical

Temperature is high all year round. Hot, humid summer where the average January maximum temperature is more than 30°C. Average absolute minimum temperature is 11°C. Experiences two seasons: 'wet season', or monsoon; and 'dry season'. Not suitable for plants that require a defined period of chilling. Extends across the north of the continent, from Rockhampton, Cairns, Darwin, to Broome and Port Hedland.

Hedges

Previous page: *Ficus benjamina* 'Exotica', *Murraya paniculata, Buxus japonica*
Above: *Buxus sempervirens*

For as long as I can remember, I've incorporated hedges into my garden designs. My ideal garden consists of strong architectural lines, including a dominant axis and maybe several secondary axes, balanced with the softness of mass planting. So hedges are a vital component, providing my gardens with strength, direction and protection. They support the garden with year-round structure. While perennials may die down and reappear, and other plants peak at certain times of the year, hedges are a constant that tie the garden together no matter what the season.

I started off my love of garden design with a strong admiration for formal Italian gardens as well as those of eighteenth-century France. On a visit to France I was inspired by the great pleached hedges of Versailles and Vaux-le-Vicomte. Although the hedges-on-stilts I was looking at were grand in scale, it occurred to me that they would be ideally suited to inner urban design as a way to save space. The clear trunk space under each tree could be used for extra planting, creating the illusion of more space. And the high hedge could still be used to obscure neighbouring buildings or fencelines. I firstly used *Pyrus* and then *Ficus* and now I experiment with many different types of plants for my hedges, including native species, citrus and other fruit trees.

Of course hedges are the very essence of a formal garden, but even the most relaxed and casual of gardens can benefit from the inclusion of a background hedge. Lilly pillies might serve as a screen to hide neighbouring buildings or an ugly paling fence. A hedge of bay might provide a perfect dark green background for mass planting in a cottage-style garden. Or a length of *Buxus* (see pages 24–25 and 27) might be allowed to imitate an old overgrown hedge, with many bumps and lumps all over the place. At my own farm I have grown the softest of hedges, with a mix of hawthorn (see page 31), rose, honeysuckle (see page 48) and *Viburnum* (see page 70), which weaves in a serpentine manner across the contours of the site.

And hedges suit any sized garden. A small courtyard can incorporate box leaf honeysuckle (see page 48) to clothe the faces of step treads and *Buxus* to neaten the edges of beds. The grandest of large country gardens can incorporate the entire spectrum of hedge styles, from

Above: *Pyrus ussuriensis*

rows of pleached hedges to parterres of small hedges.

When planning a hedge, the most important thing to consider is scale. Many people get it wrong. Some make the hedge too large and dominant for the site they are working with, dwarfing all other planting and overwhelming the other elements of the garden. The flipside is a drastically underscaled hedge, where the hedge makes little impact on the overall look of the garden and where visitors are made to feel like giants. Think carefully about the purpose of the hedge and its impact on the spaces around it before planting.

Finally, a word of warning about maintenance. It's true that hedges do require regular and dedicated care, especially if you are looking for that sleek, hard-edged look. Certain types of hedges require pruning up to three to four times a year. However, I find that even strictly formal gardens require less maintenance than elaborate flower gardens, which often need constant cutting back, dividing and staking. Consider how much effort you are willing to put in to keep it looking good before choosing the plants and hedging style for your site.

Tips for hedges

1. Resist the temptation to plant hedges too close together, as this will adversely affect the health of the plants in later life.
2. Select the correct plant for the right aspect. Consider all conditions, such as sunlight, water and competition from other trees.
3. Prune hedges from a very early age; do not wait until they get to their desired height. Regular early pruning creates a denser hedge.
4. Make sure you can obtain spare plants in case any one plant dies in the row. If supply is not guaranteed then buy extras and keep in pots in other parts of the garden so that if one is needed, it is the right size.

Acmena smithii
Lilly pilly

Mature height and spread: 4–10 m **Spacing**: 1 plant per lineal metre **Plant form/habit**: Pyramidal **Position**: Full sun to full shade **Leaf hold**: Evergreen **Benefit**: Foliage **Climate**: Cool temperate, warm temperate, subtropical **Water requirement**: Moderate

I have been using this as a very successful and extremely hardy hedge plant for the past 20 years. It is one of a few hedging plants that will tolerate deep shade as well as full sun; for this reason it is a great choice where you require uniform plants in varying conditions. Another distinct advantage is that an effective tall hedge can be maintained even when space is limited and there is not much depth to the planting area. This is handy for narrow driveways or poky side boundaries.

Although not completely drought-proof, this native Australian plant does not require much supplementary watering, and being an understorey plant it does well even with competition from larger trees. I have grown a hedge of lilly pilly very successfully right up to the base of a hundred-year-old oak tree. I have even managed to grow a decent lilly pilly hedge under a eucalyptus.

Lilly pilly does not like severe frost and for this reason it is best grown in a frost-free city or country environment. It does tolerate both dry and humid conditions. However, young hedges of lilly pilly can burn when the temperature exceeds 45 degrees for an extended period.

I find it does better in the southern part of Australia as well as in Western Australia. For Brisbane, where psyllid is a greater problem, I tend to use the slightly more psyllid-resistant varieties of *Syzygium*.

Lilly pilly is very easy to prune and will regenerate from old wood; it is, however, better if pruned regularly. I find twice to three times a year is perfect.

Although many different cultivars of this species have been developed, my preference is for the true species form which exhibits a broad and dark green leaf shape.

A new variety named 'Sublime' is proving to be an equally good plant for leaf colour and size, and this cultivar flushes lime-coloured new growth rather than red, which some people object to. At any rate, I find that even with the true species form, when developed into a mature hedge and trimmed regularly, this redness is not so prominent.

Overleaf: *Lonicera nitida, Parthenocissus tricuspidata, Prunus lusitanica, Acmena smithii*

30

Buxus japonica syn. Buxus microphylla var. japonica
Japanese box

Mature height and spread: 1–2 m **Spacing**: 4 plants per lineal metre **Plant form/habit**: Rounded **Position**: Full sun to full shade **Leaf hold**: Evergreen **Benefit**: Foliage **Climate**: Cool temperate, warm temperate, subtropical **Water requirement**: Drought tolerant

Japanese box is the toughest of all the box plants I use and will tolerate less water and more humidity than its more refined relative, the English box (see page 27). Its leaf is much larger and it gives the effect of a very robust hedge. It is a much lighter green than English box and tends to be more consistent in its colour throughout the seasons. For these reasons I use it often as a background hedge, behind more delicate hedging such as English box or box leaf honeysuckle (see page 48). Japanese box also grows faster and taller than all other *Buxus*; this usually means it requires more regular pruning, say two to three times a year.

Japanese box does not tolerate the extremely cold weather as well as English box does, but it withstands sudden summer heat better than English box. As the summer heat becomes more extreme in Victoria, for example, I'm tending to use more Japanese box when once I would have used only English box.

For smaller hedges a variety called 'Faulkner' is good. It has a more compact habit and more gloss to the foliage, and it does well in Queensland.

Buxus microphylla var. koreana
Korean box

Mature height: 300–500 mm **Mature spread**: 400–1000 mm **Spacing**: 5 plants per lineal metre **Plant form/habit**: Spreading **Position**: Full sun to full shade **Leaf hold**: Evergreen **Benefit**: Foliage **Climate**: Cold, cool temperate, warm temperate **Water requirement**: Moderate

Korean box is a variety I use when seeking a less formal hedge or for mass planting, in order to create large sheets of greenery. Its tight clusters of small leaves create a shaggy appearance and are therefore suited to more casual style. It tends to push growth outward rather than upward.

Like all other *Buxus* it is just as happy in shade as it is in sun. It does require a bit more water than its more robust cousin, the Japanese box (see opposite). It does not grow very high and is suited to either a low hedge or an area of mass planting. I find it grows well in nearly all capital cities except for Perth, where the tougher Japanese box does better in the hot, dry climate.

Buxus sempervirens
English box

Mature height and spread: 1–2 m **Spacing**: 5 plants per lineal metre **Plant form/habit**: Rounded **Position**: Full sun to full shade **Leaf hold**: Evergreen **Benefit**: Foliage **Climate**: Cold, cool temperate, warm temperate **Water requirement**: Drought tolerant

The English box plant has become synonymous with formal gardens for several very good reasons: it is generally very hardy, is easy to prune, does not require excessive trimming, and has a darkness to its green leaves like no other hedging plant. It is very dense and has a fine leaf, which makes it perfectly suited to low hedging. Because it is so slow-growing, it is usually limited to hedges up to 600 millimetres in height, which can take up to five years. Once it reaches the desired size it is one of the few hedges that can be maintained as is for centuries.

English box will not tolerate any humidity and experiences some leaf burn during intense summer heat. For these reasons it only suits the southern states and some elevated inland areas of New South Wales. It will cope with extreme cold including some snow and hard frost.

Ensure that box plants are planted into well-drained soil. If drainage is naturally poor, consider installing agricultural drainage below your garden beds. English box detests a waterlogged root zone and will protest with undesirable tones of copper and yellow appearing in the foliage.

To keep its deep green colour, feed it well with good organic fertilisers and give it lots of dolomite lime. One of my main growers recommends 2 to 3 cups of dolomite heaped around the roots every year.

For me the true beauty of English box is in its seasonal colour changes. The flush of light green new growth in the spring heralds the start of a new growing season, a time to leave winter behind and all the scarring of the previous summer. The colour of this new growth is vibrant – it brings the whole garden to life.

Overleaf: *Malus* 'Crimson Crisp', *Buxus sempervirens*

Citrus spp.
Citrus

Mature height and spread: 2–4 m **Spacing**: 1 plant per lineal 1–1.5 m **Plant form/habit**: Rounded **Position**: Full sun **Leaf hold**: Evergreen **Benefits**: Flowers, fragrance, fruit **Flowering time**: Sep–Jan **Flower colour**: White **Climate**: Cool temperate, warm temperate **Water requirement**: Moderate

Many people would never think to grow citrus plants as a hedge, but they actually make the most beautiful and aromatic of hedges. They are highly perfumed in the winter, they bear fruit and they work as a wonderful deep green screening plant. Oranges and cumquats are the best for hedging. Lemons are not dense enough to create a good structure.

I use citrus for a hedge approximately 1.5–2 metres high. They require full sun and will tolerate extreme heat in the summer. They will also tolerate cold winters and require average supplementary watering in the summer.

Citrus do not like too much humidity, and I would not recommend them for Queensland. I have, however, grown them successfully in Sydney.

Citrus hedges require minimal pruning – once a year is more than adequate. They create a loose, casual-style hedge, so don't rely on them for that perfect manicured formal look. They do reward you with beautifully scented blossoms over the spring and lots of fruit over the winter.

Oranges can be planted as close as 1.5 metres apart with cumquats at 1 metre apart.

Citrus are heavy feeders and should be fed a well-balanced fertiliser over the warm months of the year. Feeding keeps the foliage lush and deep in colour.

The best time to prune citrus is spring. Take the opportunity to prune to shape just as the new growth begins to emerge. Immediately following pruning a new flush of growth will arrive.

Scale insect and citrus gall wasp are regular pests of citrus, so be vigilant with white oil for the scale and hang yellow wasp traps within the foliage of the plant. These pests are less of an issue among strong, vigorous plants.

Crataegus spp.
Hawthorn

Mature height and spread: 2–4 m **Spacing**: 2–3 plants per lineal metre **Plant form/habit**: Rounded **Position**: Full sun **Leaf hold**: Deciduous **Benefits:** Flowers, fruit (inedible) **Flowering time**: Sep–Oct **Flower colour**: White to deep pink **Climate**: Cool temperate and warm temperate **Water requirement**: Drought tolerant

Few things represent the countryside more than a hawthorn hedge. In many older rural areas of Australia hawthorn hedges can still be seen as a form of fencing for paddocks. I always use hawthorn hedging in country gardens because it has the attraction of flowering in the spring as well as producing large fruit in the autumn. I have created a mixed hedge of all white-flowering hawthorn at my farm and this is highly successful as a tapestry hedge.

It can be kept as low as 1 metre and grows to be a large hedge up to 3–4 metres high if required. Although hawthorn grows reasonably fast it does take a while to thicken into a dense hedge.

Hawthorn trees are not always easy to source. The best time to buy them is during the winter months when they become available as bare-rooted trees; otherwise you will need try special collection nurseries at other times of the year.

In some areas certain varieties of hawthorn are a weed species, so check your local authority for advice before you plant. Keep in mind also that common hawthorn is prone to pear slug. The Washington thorn (see page 32) is my favourite among the hawthorns as it is less susceptible to pear slug.

Hawthorn requires very little supplementary water and survives despite neglect.

Crataegus phaenopyrum
Washington thorn

Mature height and spread: 3 m **Spacing**: 2–3 plants per lineal metre **Plant form/habit**: Rounded **Position**: Full sun to partial shade **Leaf hold**: Deciduous **Benefits:** Flowers, foliage **Flowering time**: Sep–Dec **Flower colour**: White **Climate**: Cool temperate, warm temperate **Water requirement**: Drought tolerant

Washington thorn is a species of hawthorn hedge that seems to be immune to the pear slug that defoliates the common hawthorn during the summer months. It makes for a great country hedge, having dense foliage and flowering prolifically in the spring. It can be maintained as a medium-height hedge of approximately 1 metre and will grow to become a large hedge up to 3–4 metres. The downside to this hedge is the thorns that cover its stems; these make pruning quite difficult and can be a risk to small children.

Washington thorn likes the same conditions as the common hawthorn, which restricts it to cooler winters and dry rather than humid summers.

× Cupressocyparis leylandii 'Leightons Green'
Leyland cypress

Mature height and spread: 20+ m, but it is very manageable to keep pruned at 3 m **Spacing**: 1 plant per lineal metre **Plant form/habit**: Pyramidal **Position**: Full sun **Leaf hold**: Evergreen **Benefit**: Foliage **Climate**: Cold, cool temperate, warm temperate **Water requirement**: Drought tolerant

Despite its reputation for being a cheap and less fashionable hedge, cypress hedging can be useful. I use it by the sea when windy, salt-laden air is a problem. Along with bay, it is one of the only hedges that will tolerate these conditions. Cypress hedging is also a great solution when extreme wind is a problem, as it works as an excellent wind barrier. I use it often in the countryside to protect ornamental or vegetable gardens. Another benefit is that it can be planted in paddocks as a windbreak with virtually no supplementary watering.

Leyland cypress grows very fast, establishing itself into a mature hedge of 3 metres within two to three years. It is always a good idea to plant it with plenty of space around it. The small plants will overtake larger trees in no time.

Leyland cypress will tolerate both heat and cold, as well as dry and humid conditions, although it does prefer less humidity.

It will require regular pruning due to its fast growth habit, say two to three times a year. Its roots can be invasive to the garden, so plant them a good distance away from your garden beds and install a root-control barrier between the bed and the hedge.

The major downside of cypress hedging is the increasing prevalence of canker – a disease that causes dead brown patches within the hedge. If it sets in it can take over the whole hedge and destroy it. A cure for this is near impossible so act to prevent this disease by ensuring excellent drainage.

Cupressus macrocarpa
Monterey cypress

Mature height: 20–30 m **Mature spread:** 10–15 m **Spacing:** 1 plant per hedge width, eg if hedge is 1 m wide = plant every 1 lineal metre **Plant form/habit:** Cone-shaped as juvenile tree, then spreading with maturity **Position:** Full sun **Leaf hold:** Evergreen **Benefits:** Foliage for visual screening and shelter **Climate:** Cool temperate, warm temperate **Water requirement:** Drought tolerant

Whenever I am in need of a very robust, tall, wind-tolerant hedge, my first choice would be a cypress hedge. The most common of cypress hedges is the Leightons Green; it's fast-growing and very tough, but for a tighter and more exact cypress hedge I use *C. macrocarpa*. Due to its almost needle-like leaf you can, with constant clipping, keep this hedge very dense and precise; I have even seen the name of a house cut into it. It tolerates extreme heat and high winds and requires very little watering in the summer months. It was used extensively throughout the western farming districts of Victoria as a windbreak. Good in coastal gardens as it's tolerant of salt spray and wind. Soil must be well drained, to promote vigour and disease resistance. In poorly drained soils, trees become weakened and are susceptible to fungal cypress canker. A preventative copper-based spray program may be necessary in canker-prone areas.
I find if you keep the plant very healthy, well-watered and drained, it is far more resilient against this disease. It is slower growing than Leightons Green but well worth the extra waiting time.

Ficus carica
Fruiting fig

Mature height and spread: 2–4 m **Spacing**: 1 plant per lineal metre **Plant form/habit**: Rounded **Position**: Full sun **Leaf hold**: Deciduous **Benefits**: Foliage, fruit **Climate**: Cool temperate, warm temperate **Water requirement**: Drought tolerant

I first saw a fruiting fig hedge at the base of the National Portrait Gallery in London and completely fell in love with its highly sculptural qualities. It was the perfect heavy texture to visually support the monumental weight of that building. I had never seen fig used this way before and have never since, but it appealed to me immensely. I don't use it very often, saving it for special situations and for people who appreciate unusual qualities in a garden. It is very slow to establish and often difficult, as fig in its younger years can be a little temperamental. But after two to three years of patience, it rewards you with a highly prized hedge that everybody responds to.

Fig seems to grow in most parts of Australia, tolerating both humid and dry conditions. It does not like the extreme cold, however, with heavy frost causing the limbs to die back.

When establishing your hedge be patient, buy the plants small and cut them back regularly to maintain as tight a hedge as you can, remembering you are growing it for its texture and not density.

Fruiting figs are not always available in pots through the summer, but can be ordered in the autumn and supplied as bare-rooted trees for planting in the winter months. If you are hunting for fruiting figs in pots, you may need to seek out nurseries with a special emphasis on fruit-producing plants.

Ficus microcarpa var. hillii
Hill's fig

Mature height and spread: 4–10 m **Spacing**: 1 plant per lineal 2–3 m **Plant form/habit**: Rounded **Position**: Full sun to partial shade **Leaf hold**: Evergreen **Benefit**: Foliage **Climate**: Warm temperate, subtropical and tropical **Water requirement**: Moderate

Hill's fig is my favourite and most commonly used tall hedge. It seems to tolerate nearly all conditions except for snow and extreme frost. I use it successfully in all capital cities of Australia, including the dry heat of Perth right through to the humid conditions of Brisbane.

It responds well to pruning and for that reason makes an ideal screening hedge. I pleach it often as a boundary hedge, where it will clip into the perfect squared-up hedge screen.

Hill's fig develops a lovely smooth white bark which adds to its appeal as a pleached hedge. Select trees with a nice straight single leader and trim away branches lower than 1.1 metres to make this attractive trunk more pronounced.

Possums love the soft new foliage and can cause much damage to a juvenile hedge. Do not panic at this stage as the overall hedge will outgrow the possums to form a dense mature hedge.

One of the problems of a *Ficus* hedge can be their very fibrous, aggressive root systems. For this reason I recommend a root-control barrier to be used to protect neighbouring properties and I would not plant them too close to a house. There is a slightly more compact variety of *Ficus* that is more suitable to smaller properties and hedges – the *Ficus hillii* 'Flash'. I often use this when space is more limited and such a tall hedge is not required.

As with most large trees used for hedging, if the Hill's fig is trimmed regularly the root system problem is greatly diminished. The other side effect of the fibrous root system is the difficulty in growing other plants under the hedge. Over time the soil tends to be saturated with *Ficus* roots, making it difficult for other plants to survive. Two plants I have found that do co-exist happily with the *Ficus* are *Murraya paniculata* (see page 52) and *Trachelospermum jasminoides* (see page 200).

Laurus nobilis
Bay laurel

Mature height and spread: 4–10 m **Spacing**: 1 plant per lineal metre **Plant form/habit**: Upright **Position**: Full sun to full shade **Leaf hold**: Evergreen **Benefits**: Flowers, foliage, culinary herb **Flowering time**: Sep–Nov **Flower colour**: Pale yellow **Climate**: Cold, cool temperate, warm temperate **Water requirement**: Drought tolerant

One of the oldest and most reliable hedges is the bay hedge, used by the Romans extensively. Nothing characterises Mediterranean gardens more than a bay hedge. I love it for its aromatic and very dark leaves and I use it for clients who are willing to be patient and wait a few years before they get the full, glorious benefit of this plant. They also bear small pale yellow flowers in the spring.

Being southern European the bay laurel does well in dry hot summers. It will only just tolerate the humidity of Sydney, but does not cope with that of Brisbane. It will survive frost and snow but not wet feet, preferring a well-drained soil.

It is possible to keep it clipped as a medium hedge at a height of 1 metre, or let it reach a height of over 4 metres. It is not fast growing and may take four to five years before thickening into a dense hedge. But once it is established, it will reward you by requiring very little maintenance and very little water.

Bay is not suited to pleaching due to its irregular trunk shape.

One of the misconceptions about bay hedges is that they require full sun. This is not true. I have grown them very successfully in partial and even full shade.

Another great attribute of bay is that it can thrive under established trees. I grew a very successful hedge immediately under a long-established poplar tree using very little water.

Bays have a tendency to send out suckers, which is a good attribute for a hedge. However, it is a wise idea to install a root barrier beside the bay if you intend to nurture a garden plot beside it. This will prevent woody suckers entering valuable soil space.

Scale insects can sometimes affect bay trees. They can be hard to spot, but a sure indicator is the presence of ants wandering up and down the stems, feeding on the honeydew left behind by the scale. When you see the ants, one simple spray of white oil will eradicate the scale insect altogether.

Ligustrum vulgare
Privet

Mature height and spread: 2–4 m **Spacing**: 3 plants per lineal metre **Plant form/habit**: Rounded **Position**: Full sun **Leaf hold**: Evergreen **Benefit**: Foliage **Climate**: Cool temperate, warm temperate **Water requirement**: Drought tolerant

When I first worked in the Hamptons in America I discovered the marvels of the privet hedge. There it is used exclusively as a boundary planting and replaces all fences. It is used for good reasons, as it tolerates the extreme conditions of New York State: frost, snow, humidity and heat do not seem to hurt the plant at all. It also prunes extremely well and will regenerate from old wood, thus allowing you to halve the height of your hedge if you wish. My main pruning contractor describes pruning privet as like slicing butter with a hot knife.

Some authorities list privets as weeds, although not usually the *Ligustrum vulgare*. But check with your local authority before planting.

The real advantage to privet hedge is that it is fast growing and very tough, with virtually no pests or diseases, and requires very little water. It will also do well in windy conditions, but will not tolerate wet feet.

I have not seen it growing in humid states and would therefore suggest it is limited in Australia to areas south of Queensland.

I have used privet at both my country properties with much success, where they are given a major prune once a year and several lighter tidy-up trims. It feels to me more like a country-style hedge and is great where large areas of hedging are necessary, due to its rapid growth habit.

Privet can be semi-deciduous in really cold weather, although I have found that once it is established as a mature hedge it is less deciduous in nature. It may also turn a slightly purple colour in the winter.

Be warned that it is known for its rather foul-smelling flowers, to which some people may be allergic. With regular pruning the hedge ceases to flower.

Privet is not commonly available in garden centres. It is a plant that you will have to ask for and your local nursery should be able to place special orders for hedging quantities.

Lonicera nitida
Box leaf honeysuckle

Mature height and spread: 1–2 m **Spacing**: 7–8 plants per lineal metre **Plant form/habit**: Rounded **Position**: Full sun **Leaf hold**: Evergreen **Benefit**: Foliage **Climate**: Cool temperate, warm temperate **Water requirement**: Moderate

I first saw box leaf honeysuckle hedged at Hidcote in England, where it was used to clothe the risers on a set of steps. I have since successfully replicated this technique in Australia, and it has translated very well in this country in nearly all areas. This is the way I use this plant. As a normal hedge it requires far too much trimming and very quickly grows out of shape.

It has a wonderful ability to live in the small pocket of soil trapped in each step tread. Do not overwater or overfeed it, and trim it regularly.

Murraya paniculata
Orange jessamine, mock orange

Mature height and spread: 2–4 m **Spacing**: 2–3 plants per lineal metre for low hedge; 1 per lineal metre for larger hedge **Plant form/habit**: Rounded **Position**: Full sun to full shade **Leaf hold**: Evergreen **Benefits**: Flowers, foliage, fragrance **Flowering time**: Oct–Apr **Flower colour**: White **Climate**: Warm temperate, subtropical, tropical **Water requirement**: Moderate

Orange jessamine has proven to be the toughest of all hedges, surviving nearly all climatic conditions. It tolerates some frost, just not black frost (an extreme frost without snow that blackens the vegetation). This wonderful plant even tolerates a large degree of root competition from other trees. I plant it under my *Ficus hillii* hedge (see page 41) and it seems to thrive here. The only condition it will not tolerate is too much water – wet feet cause it to yellow and die.

Its texture and colour complement that of the Japanese box (see page 24) and I often layer these two hedges together. When using a double hedge, I make the *Murraya* the larger of the two hedges and therefore at the rear.

Murraya paniculata forms a fast-growing hedge which requires regular trimming, less in the southern states and far more in the northern, more humid states, where the plant absolutely thrives. It can be kept as low as 300 millimetres and will grow as high as 2–4 metres. It benefits from being treated as a formal hedge or informal screen, because the regular clipping generates thick bushy growth and without this it can open up to be a regular woody shrub with less visual impact.

Murraya paniculata boasts dark green, glossy foliage, and for this alone, it has immediate appeal. Regular compost and fertiliser applied to this plant over the warmer months will help to maintain this deep foliage colour through the winter.

One of the attractions of *Murraya paniculata* is its strong perfume. Unfortunately when pruned as a hedge it does not flower – the pruning removes all the flowers. If you want some flowers you could let the hedge flower and then prune hard after flowering; this will leave you with a rather loose, informal hedge for part of the year.

I have grown *Murraya paniculata* under trees reasonably well, and it is one of those rare hedges that will tolerate both sun and shade, making it very handy where consistency is required in varying light levels. Keep in mind, though, that it does prefer more sun than shade.

I don't believe it is as well suited to country properties as to city environments, so I tend to use it only in urban settings.

Some authorities list *Murraya paniculata* as weedy, so check with your local authority before planting.

Osmanthus heterophyllus 'Purpureus'
Holly leaf tea olive

Mature height: 4 m **Mature height**: 2 m **Spacing**: 700 mm off-centre **Plant form/habit**: Bushy **Position**: Full sun to partial shade **Leaf hold**: Evergreen **Benefits**: Foliage **Flowering time**: Feb–May (late summer to autumn) **Flower colour**: White **Climate**: Cold, cool temperate **Water requirement**: Drought tolerant

This *Osmanthus* is a great climate change hedge alternative. It is extremely drought hardy and tolerates sustained extreme temperatures. It also requires minimal trimming, a very important requirement in the constant search for low maintenance gardens. If you are looking for the hedge with the least amount of pruning, this is the one. Its evergreen leaves are dark glossy green.

It's not a fast-growing hedge or a tall one – use this plant if you require a medium hedge to under 2 metres. It is spiky, so use with care around children or sensitive gardeners. It's very dense and provides a good barrier for boundaries.

In the growing season its new growth is a beautiful deep purple colour, the leaf form resembling a holly leaf. Its flowers are small and white with a slight perfume.

Prunus laurocerasus
Cherry laurel

Mature height and spread: 4–10 m **Spacing**: 1 plant per lineal metre **Plant form/habit**: Rounded **Position**: Full sun to full shade **Leaf hold**: Evergreen **Benefits:** Flowers, foliage **Flowering time**: Sep–Nov **Flower colour**: Cream **Climate**: Cold, cool temperate, warm temperate **Water requirement**: Moderate

For large country gardens, especially ones that do not have a lot of water, the cherry laurel plant makes a great hedge. It thrives in cold mountain conditions and will also take extreme summer heat. It seems to dislike the extreme humidity of northern states, though it does tolerate some.

It is a relatively fast-growing hedge plant – it will grow to a couple of metres in two to three years. It also bears showy cream-coloured flower spikes in the spring, but it is grown more for its foliage interest.

Its leaf is a long, glossy light green, fading to darker green in the summer. Because of its long leaves, it does not create the densest of hedges, so it suits a looser, less formal style. For some reason I associate it with hill gardens, probably as it tolerates the wet conditions that mountains provide.

It is also worthwhile looking at the baby cherry laurel, 'Otto Luyken'. This plant requires the same environment for growth but can be maintained at less than 1 metre in height and has a smaller leaf in both width and length. The foliage is once again a desirable dark glossy green and is sometimes used in floral arrangements.

Juvenile plants of the cherry laurel can exhibit small holes in the foliage. This is a fungal condition known as 'shot hole' disease. Don't be put off by this early appearance, because once the plants establish themselves into natural soil, the vigour of the plant suppresses this condition and you would not know the plant ever had the disease.

Prunus lusitanica
Portuguese laurel

Mature height and spread: 4–10 m **Spacing**: 1 plant per lineal metre **Plant form/habit**: Rounded **Position**: Full sun to full shade **Leaf hold**: Evergreen **Benefits**: Flowers, foliage **Flowering time**: Sep–Nov **Flower colour**: Cream **Climate**: Cold, cool temperate, warm temperate **Water requirement**: Moderate

This is one of my favourite medium-sized hedges. I use it in both city and country locations at about 1.5 metres tall. I love it for its rich, dark, flat green leaves. It works so well as a strong background to flowering perennial borders.

It is a little temperamental to get established for some reason I cannot fathom, but once established it is very tough. It forms a dense hedge that can range from 1 metre to 4–5 metres. It has beautiful flowers that remind me of the white racemes of wisteria. Keep in mind, though, that like many plants, once trimmed into a hedge it does not flower.

I find it really only needs trimming once a year or, for a loose hedge, once every two years.

Every spring you can look forward to a wonderful flush of new growth, which is very rewarding. It is also appealing to caterpillars so be vigilant during this time and pounce on any caterpillars with a low-hazard spray before they affect the foliage for the entire season.

My only worry with this plant is the fine balance it requires with its watering regime – too much early on kills it very quickly, while lack of water also adversely affects it. I would use less water rather than more when planting Portuguese laurel as it seems to prefer drier conditions; for this reason it is not suitable for wet mountain regions.

The Portuguese laurel will tolerate the cold, surviving both frost and snow. It is suited to dry summers and does not do well in the humidity of the northern states. It tolerates summer heat, although heatwaves up near the 45°C mark will burn its foliage.

Opposite: *Prunus lusitanica*, *Euphorbia* sp., *Prunus* specimen tree

Pyrus spp.
Ornamental pear

Mature height and spread: 4–10 m **Spacing**: 1 plant per lineal 2.5–3 m **Plant form/habit**: Rounded **Position**: Full sun **Leaf hold**: Deciduous **Benefits**: Flowers, foliage, fruit (inedible) **Flowering time**: Aug–Sep **Flower colour**: White **Climate**: Cold, cool temperate, warm temperate **Water requirement**: Drought tolerant

Pyrus hedges have become one of my signature styles. I use them for their form, colour, flowers and most importantly their extreme toughness. I first discovered them being used in the hot summers of Texas, in the USA, and knew they would be suitable for Australia.

I first started using the Manchurian pear (see pages 332–33), for its ability to withstand pleaching. However, it is not a good specimen tree as it is prone to splitting. Since I first used *Pyrus* for hedging, many more suitable varieties have become available, mostly belonging to the *Pyrus calleryana* species. These varieties include 'Chanticleer', 'Bradford' and 'Red Spire'. They have a better branch structure and are narrower in form, making them ideal for inner-city hedges.

The other major advantage of *Pyrus* hedges is they have one of the shortest deciduous periods; in fact, in milder climates such as Sydney they can hold their leaves well into June. You are rewarded with deep crimson colour in the autumn, followed by a profusion of white blossom in the spring. And for their size they also have a relatively non-aggresive root system. Although all trees have the potential to damage nearby buildings, *Pyrus* are one of the safest choices when planting close to a structure. It's hard to find a problem with the *Pyrus* genus!

Pears respond well to pruning, with *Pyrus ussuriensis* shaping into a dense square pleached hedge, while the *P. calleryana* species shape more into a narrow, rectangular pleached hedge.

They can be bought as fairly advanced trees but I think they are better planted as trees 1–2 metres in height and trained from an early age to hedge. This makes for a tighter, more controlled hedge.

They really are the all-in-one package tree: seasonal features, excellent tolerances and pest- and disease-free. They are hard to fault.

Pyrus calleryana
Callery pear

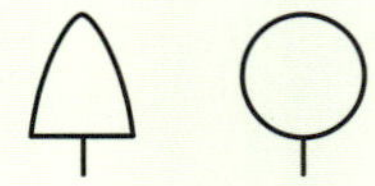

Mature height: 10–15 m **Mature spread:** 3–9 m **Spacing:** 1.5–2 m apart **Plant form/habit:** Pyramidal, becoming rounded with maturity **Position:** Full sun **Leaf hold:** Deciduous **Benefits:** Flowers, foliage **Flowering time:** September **Flower colour:** White **Climate:** Cold, cool temperate, warm temperate **Water requirement:** Drought tolerant

The benefits of this tree are well outlined in the tree section of this book, but it is worth mentioning in the hedge section as I mostly use it this way. *Pyrus* trees respond extremely well to pruning and shaping, with much lateral growth, making them ideal pleaching trees. I plant them 2–3 metres apart when using as a hedge, the further apart the better for the health of the tree. It's better to start with a smaller tree of approximately 1.5 metres high and one with a very straight trunk. As the tree grows, strip off all lower foliage and branches to 1.2–1.4 metres high and then allow the foliage above this height to merge with the tree adjacent. Regularly prune the sides and the top of the foliage until you reach the desired height. I like to keep the width of the foliage to approximately 1–1.2 metres. Once you have the hedge at the size and height you require, it will only need pruning once to twice a year. This hedge will reward you with white spring blossom as well as red and orange tones to the autumn foliage. *P. calleryana* loses its leaves for the shortest period during winter out of all the *Pyrus* trees. It makes for an ideal inner-city pleached hedge.

Pyrus ussuriensis
Manchurian pear

Mature height: 9 m **Mature spread:** 7 m **Spacing:** 2 m apart **Plant form/habit:** Rounded **Position:** Full sun **Leaf hold:** Deciduous **Benefits:** Flowers, foliage **Flowering time:** Aug–Sep **Flower colour:** White **Climate:** Cold, cool temperate, warm temperate **Water requirement:** Drought tolerant

This wider growing *Pyrus* tree also makes for a good pleached hedge. As a tree it's more susceptible to branch splitting in high winds, but as a hedge it tends to be more robust. Its advantage over *P. calleryana* (opposite page) is the more intense red autumn tones, and its potential to create a larger hedge. It will lose its leaves for a longer period than its cousin and is harder to contain as a narrow hedge, which is often important to inner-city gardeners. It has a darker leaf and presents better white blossom in the late winter to early spring. Train it as a hedge as for *P. calleryana*. Prune to promote horizontal branching from the main leader.

Syzygium australe
Brush cherry

Mature height and spread: 4–10 m **Spacing**: 1 plant per lineal 1.5 m **Plant form/habit**: Rounded **Position**: Full sun to partial shade **Leaf hold**: Evergreen **Benefit**: Foliage **Climate**: Cool temperate, warm temperate, subtropical, tropical **Water requirement**: Moderate

The brush cherry is closely related to the lilly pilly (see page 21) and enjoys very similar growing conditions. When and how I use brush cherries rather than lilly pillies comes down to a few subtle factors.

Brush cherries do not do as well as lilly pillies when planted under established trees – the cherry does not like the competition. The leaf of the brush cherry is glossier and slightly rounded, so it makes a better stand-alone hedge. I use lilly pillies more often as a background hedge.

Both brush cherries and lilly pillies can be kept pretty narrow. I have used both successfully in tight situations between a side boundary and drive where only 400 millimetres may be available.

I do find that the brush cherry requires slightly higher light levels than the lilly pilly, so if the area you want to hedge is deeply shaded, use lilly pillies.

Both lilly pillies and brush cherries are prone to psyllids and a new leaf-eating beetle, with brush cherries being more susceptible. New cultivars are constantly being selected for their ability to resist psyllid insects. One variety I have been using successfully is 'Resilience'. I like its rounded glossy dark leaf and, as its name suggests, it resists psyllid, though does not always escape attack. In my experience, I have found that once brush cherry hedges are well established, they are less susceptible to the psyllid pest.

Another variety that has been proving itself a good performer is 'Hinterland Gold'. It is slow to establish, but it has a very compact habit and requires little if any clipping to create a private urban foliage screen. Most brush cherries exhibit dark green foliage with red-tinged new growth, while the 'Hinterland Gold' sprouts pale green new foliage which then matures into mid green.

Tilia sp.
Linden, lime

Mature height and spread: 10+ m **Spacing**: 1 plant per lineal 3 m **Plant form/habit**: Upright **Position**: Full sun to partial shade **Leaf hold**: Deciduous **Benefit**: Foliage **Climate**: Cold, cool temperate **Water requirement**: Moderate

There is nothing more stunningly European than a pleached lime hedge, with uniform, thick trunks and light green foliage. They do grow well in some regions of Australia, given the right conditions, and as long as the summer is not too hot. I have seen very healthy pleached lime hedges in the Southern Highlands of New South Wales, in Tasmania, and in the more elevated mountainous regions of Victoria and New South Wales, where the summer temperature is low. They do not tolerate extreme heat in the summer and will suffer from leaf scorch. This seems to be a problem chiefly when the plant is juvenile and diminishes as it matures.

Lindens are deciduous. This may be a disadvantage if you are trying to screen out neighbours, but as a stand-alone hedge in the middle of the garden, their branch structure in the winter is highly sculptural and almost as beautiful as when in full leaf. The light green colour of the heart-shaped leaves attracts attention in the garden instead of receding as most hedges do.

Lindens are mostly used for specialised cold-climate gardens in Australia, and are not commonly available. They do not establish quickly, but if you have the right conditions this hedge is well worth the wait. Use it as a major feature in the garden.

When grown as a pleached hedge the clear trunk height should be approximately 1.2–1.4 metres. The overall height including foliage can then be 3–4 metres.

Due to its delicate nature, give the plant ideal conditions – good supplementary water, free drainage and minimal competition from surrounding trees. It will require regular trimming and in its early years will benefit from being trained strictly on a wire and post frame.

The most popular variety is *T. cordata* 'Greenspire' but my favourite is *T.* × *europaea* for its large leaf (see page 346).

Viburnum tinus
Laurustinus

Mature height and spread: 2 m **Spacing**: 1 plant per lineal metre **Plant form/habit**: Rounded **Position**: Full sun to partial shade **Leaf hold**: Evergreen **Benefits**: Flowers, foliage, fragrance **Flowering time**: Oct–Nov **Flower colour**: White **Climate**: Cool temperate, warm temperate **Water requirement**: Drought tolerant

Viburnum tinus has been used as a hedge in Australia for many years now and has gone in and out of fashion many times. I see it as an easy hedge to plant without being terribly special.

It falls into the 'tough as old boots' category. I have even seen it growing under old, established pine trees, which usually make conditions impossible for other plants to thrive. It provides a good dense hedge to 2 metres high with dark matt green foliage. If time allows between prunings, it will also produce white flower heads.

This plant thrives on neglect and has proven itself to survive drought and cope with hot summers. But if it is treated to an irrigated garden, moisture and humidity are sufficient deterrents for the red spider mite to which *Viburnum tinus* is very susceptible. I have seen whole hedges defoliated by the bug. This can be sprayed with an insecticide, but in this era of minimal chemical usage I usually rule it out. If red spider mite is not prevalent in your area, you don't have abundant water supplies and there is some degree of competition from other trees, then this hedge may be for you.

Due to the size of the leaf, *Viburnum tinus* does not clip to form a highly precise hedge, so I recommend it for use in less formal, country-style gardens. The colour is dark green in summer and light green in the spring. One benefit is that it will still flower when clipped as a hedge, again adding to the country feel of this plant.

Viburnum tinus is restricted to areas of low humidity as it is prone to sooty mould in more tropical environments. If I use it, I restrict it to Victoria, South Australia, Tasmania and parts of southern rural New South Wales.

Waterhousea floribunda
Weeping lilly pilly

Mature height and spread: 4–10 m **Spacing**: 1 plant per lineal 2 m **Plant form/habit**: Rounded **Position**: Full sun to partial shade **Leaf hold**: Evergreen **Benefit**: Foliage **Climate**: Warm temperate, subtropical, tropical **Water requirement**: Moderate

Waterhousea has become one of my favourite plants for large hedges and I now use it extensively. It is native to Australia, which makes it highly suitable as a landscape plant in this country.

This tall dark green tree can grow very large, but when kept as a hedge its ideal range is anywhere from 2 metres to 7 or 8 metres. It needs a bed width of 1 metre or more to support a good canopy height. I use it as a hedge to the ground and do not pleach it, as its trunks tend to be too irregular.

As its common name suggests, this lilly pilly does 'weep', with its leaves always pointed downwards, making for a less formal hedge that moves gracefully in the wind. For this reason it is suitable for a more organic, contemporary-style hedge. It is a hedge that does not require excessive trimming; once every two years is enough to keep it in check. If you prefer a more formal hedge it can be trimmed annually or twice a year.

At the beginning of every growing season weeping lilly pillies put on a flush of new growth, starting out red-tinged then turning lime-green and eventually hardening off to dark green through the summer months. This makes them seasonally interesting.

The other big advantage is that specimens are available as established trees. So it is possible to have an instant hedge up to 5 metres.

Waterhousea prefers full sun and thrives in this situation. It has been known to burn in extreme summer heat in the southern states. It will tolerate both dry and humid summers, making its use widespread throughout Australia. I have grown it in shade but it does not like full shade and does not tolerate root competition from large surrounding trees. I find *Acmena* lilly pillies are a better alternative in this situation.

Weeping lilly pillies are prone to psyllids as juvenile plants, but after they are established the trees offer enough natural defences to fend off this pest. I have found that they are less susceptible to psyllid attack than other lilly pillies (see page 21) and the brush cherry (see page 68).

One warning: there can be a lot of variability with the quality of weeping lilly pilly specimens. It is not always best to go for the tallest plants; rather, select trees that have been clipped through their nursery lives. This will be evident in good bushy specimens, where branching is well covered by the foliage.

Clouding

Previous page: *Ligustrum vulgare* 'Buxifolium'
Above: *Lavandula dentata*

Much has changed in landscaping styles over the past 10 years. Today, the trend is towards less formal, softer and wilder gardens.

As a result, the manner of shaping hedging material has also changed. Straight lines are no longer fashionable and more organic forms and layouts prevail. The formal parterre, a much-loved section of the garden at Stonefields, would now be considered too formal. I have long thought about removing this part of the garden and replacing it with large swathes of flowering perennials and defining the corners with clipped, organic shapes of box. But the thought of removing all this planting, which has taken over a decade to mature and countless days of rigorous and precise pruning every year to maintain, always prevents me from proceeding.

What I have been introducing into other parts of the garden (in particular, the farmhouse garden) and in designs for clients are lots of cloud-pruned organic shapes. Clouding is formed by regularly clipping plants from a young age into forms that resemble clouds. In doing so, they often mould into the next cloud, thus forming long drifts of these forms. The effect is waves of soft clipped forms that can dominate an entire garden bed or simply define corners of accent points in the garden.

Clouding can be created from many different types of plants listed here in this chapter, some native and some exotic species. It provides softness to the garden but nevertheless a strong sense of form that would previously have been created using formal, straight lines of hedges. I particularly like using them in the corners of garden beds to soften right angles, or in the bends of long curvaceous garden beds, to add a sense of formality.

I also love contrasting clouding with flowering perennials, allowing spilling perennials such as salvias

or penstemons to create a backdrop or even a low foreground to their placement. Cloud shapes seem to work better in groupings; in particular, one large specimen flanked by two smaller ones, as a single sphere can tend to look lonely and more rigid in the garden. They can consist of any shape you desire and are often formed naturally by the hand of the pruner or the growing habit of the plant. For a slightly more formal look, use perfectly round spheres in clusters. An ideal example of this is the apple walk at Stonefields which consists of large lengths of different-sized English box spheres.

I think the important thing with clouding is to have fun with it. Don't treat the design too seriously; create soft clouds that merge with each other in forms that resonate with your personality.

Tips for clouding

1. Make sure you select the correct plant for the climatic conditions – in particular, shade and sun requirements. Clouding can be taxing on a plant and needs optimal growing conditions to work.
2. Work with the growth form of the plant. Don't try and contort it too much. When clipping, be guided by the natural dips and curves of the plant.
3. Work out how much maintenance you want to do. Some plants such as *Rhaphiolepis* (see page 80) have natural clouding forms that require little clipping, while some like *Ligustrum vulgare* 'Buxifolium' (see page 79) require more regular pruning.

Buxus japonica syn. Buxus microphylla var. japonica
Japanese box

Mature height and spread: 1–2 m **Spacing**: 4 plants per lineal metre **Plant form/habit**: Rounded **Position**: Full sun to full shade **Leaf hold**: Evergreen **Benefit**: Foliage **Climate**: Cool temperate, warm temperate, subtropical **Water requirement**: Drought tolerant

I have been using *Buxus* in many forms, including in edging for beds, elaborate parterres and single spheres in corners of beds, throughout my entire career. Recently, my preferred species *Buxus sempervirens* (see page 27) came under attack from a mite that weakens the plant and eventually kills it. I have changed to Japanese box for this reason, as it appears to be less prone to mite attack and other diseases.

Buxus clouding is one of the great forms of contemporary garden design used throughout the world. The huge tolerance *Buxus* has to temperature ranges and its soft foliage that clips easily into myriad shapes and sizes makes it ideal for clouding. It will thrive in shade and full sun, as well as humid or dry conditions. The dark-green, glossy leaf provides interest throughout the year, with a flush of light green in the spring providing seasonal attraction. Prune it once to twice a year to keep the cloud form and feed it regularly during the growing season.

Correa alba
White correa

Mature height and spread: 1.5 m **Spacing**: 700 mm off-centre **Plant form/habit**: Rounded, spreading **Position**: Full sun to partial shade **Leaf hold**: Evergreen **Benefits**: Foliage, drought-tolerant, coastal plant **Flowering time**: Mar–Aug (autumn and winter) **Flower colour**: White **Climate**: Cold, cool temperate, warm temperate **Water requirement**: Drought tolerant

Correa is similar to *Westringia* (see page 81) in its growing habit and colouring, with some types slightly greyer than the *Westringia*. Both being Australian native plants, they sit well together, and I often use them in combination.

As with *Westringia*, *Correa* does extremely well in sandy soils and is suited to coastal settings. It does need light irrigation and appreciates some compost such as mulch. It needs to be trimmed from an early age to prevent pruning into thicker stems. Trim regularly and ensure it has good drainage.

Lavandula dentata
French lavender

Mature height and spread: 500–1000 mm **Spacing**: 4 plants per lineal metre **Plant form/habit**: Rounded **Position**: Full sun **Leaf hold**: Evergreen **Benefits**: Flowers, foliage, fragrance **Flowering time**: Year-round **Flower colour**: Purple **Climate**: Cool temperate, warm temperate **Water requirement**: Drought tolerant

There are many types of lavender I use in my designs, but French lavender or *Lavandula dentata* is the only one I recommend for clouding. It is longer living than the English and Italian lavenders when pruned and forms a much tighter foliage form. In saying that, lavenders are all very temperamental and prone to dying. It's not uncommon to have an entire row of lavenders where just one dies while the others all thrive. They require excellent drainage, hating any form of wet feet.

French lavender needs regular pruning (up to three to four times a year) to maintain its shape. I often allow it to flower for a few months and then cut it back to its cloud form. This can be done up to twice a year. French lavender doesn't tolerate hard frost as well as its English cousin, but loves hot, dry conditions with no humidity. It grows well in coastal conditions as well as well-drained, rich soils.

Ligustrum vulgare 'Buxifolium'
European privet

Mature height and spread: 2–4 m **Spacing**: 3 plants per lineal metre **Plant form/habit**: Rounded **Position**: Full sun **Leaf hold**: Evergreen **Benefit**: Foliage **Climate**: Cool temperate, warm temperate **Water requirement**: Drought tolerant

Ligustrum's very tight, small leaf form makes it an excellent choice for clouding. It has none of the mite or disease issues plaguing *Buxus* (see opposite), and once clipped closely resembles *Buxus* in appearance. It grows much faster, too, reaching the desired shape more quickly. But this also means it needs clipping more regularly, up to three to four times a year.

Some councils have banned all *Ligustrum* as weeds, so check with your local authority before deciding to use it. It may be hard to locate due to its relatively little-known exposure in the Australian market. It thrives on heat, frost, some humidity and low water use.

Pittosporum tobira 'Miss Muffet'
Japanese pittosporum

Mature height and spread: 2–4 m **Spacing**: 2 plants per m^2 **Plant form/habit**: Rounded **Position**: Full sun to full shade **Leaf hold**: Evergreen **Benefits**: Flowers, foliage, fragrance **Flowering time**: Jan–Mar **Flower colour**: Cream **Climate**: Cool temperate, warm temperate, subtropical **Water requirement**: Drought tolerant

This variety of *Pittosporum tobira* is readily available in Australia and is popular due to its compact form. Only use it in full sun or very light shade; if it receives too much shade the stems will elongate and destroy its compact nature. It requires very little pruning, if any, to maintain a clouding form and will also reward you with scented cream flowers in summer.

The 'Miss Muffet' variety of *Pittosporum* is great in coastal conditions where it benefits from sea breezes. It is prone to attack from insects and some diseases, so always grow it in a well-ventilated position.

Rhaphiolepis × indica
Indian hawthorn

Mature height and spread: 1 m **Spacing**: 1 plant per m^2, or 700 mm apart if planted as a border **Plant form/habit**: Rounded **Position**: Full sun to full shade **Leaf hold**: Evergreen **Benefits**: Flowers, compact shrub form, dark glossy green foliage **Flowering time**: Early autumn to late spring **Flower colour**: White flowers with pink stamens, gives blush appearance **Climate**: Cool temperate, warm temperate, subtropical **Water requirement**: Drought tolerant

Rhaphiolepis, with its shiny green foliage and neat mounding form, is perfect for clouding. Both the 'Oriental Pearl' and 'Spring Pearl' varieties make for a good clipped shape; you can often leave them for a few years without pruning and they will maintain a clouding form. Mix them with either the complementary colour of *Buxus* (see page 78) or the contrasting foliages of *Westringia* (see opposite) and *Correa* (see page 78) for a more dramatic effect. Once it forms the desired shape it needs very little pruning – up to once a year.

Rhaphiolepis is drought hardy and will tolerate both extreme heat and cold, as well as some humidity. Another advantage is its flush of pale pink-to-white flowers, which cover the entire plant. This is one of my go-to plants for clouding due to its extreme hardiness. It doesn't like too much shade but will tolerate competition from neighbouring trees. It is considered a weed in some areas.

Westringia fruticosa
Coast rosemary

Mature height and spread: 1–2 m **Spacing**: 1 plant per lineal 600 mm or 2 plants per m^2 for a massed-planting effect **Plant form/habit**: Rounded **Position**: Full sun to partial shade **Leaf hold**: Evergreen **Benefits**: Flowers, foliage **Flowering time**: Sep–Nov **Flower colour**: White **Climate**: Cold, cool temperate, warm temperate **Water requirement**: Drought tolerant

Westringia has become one of the most popular plants for clouding. Its drought hardiness and ability to grow in coastal conditions make it the perfect plant for this form of pruning. Its colour contrasts well with the bright green shine of other clouding plants such as *Rhaphiolepis* and *Buxus* and complements the sage-green native colourings of *Correa* and other native plants.

I use *Westringia* more often in a coastal setting, where it suits sandy soils and salt wind. It grows well in richer soils, but needs good drainage.

Westringia will not tolerate extreme frost and high humidity. Minimal clipping (up to twice a year) is required to maintain its shape. It suits light shade but prefers full sun.

Shrubs

Previous page: × *Cupressocyparis leylandii* 'Leightons Green', *Lavandula angustifolia, Santolina chamaecyparissus*
Above: *Buxus sempervirens, Hydrangea macrophylla*

If trees and hedges form the structure of a garden, then shrubs are the filling required to make it all look like a garden. Shrubs provide substance and are often used as a backdrop to smaller feature plants. Without shrubs the garden looks incomplete – they are a vital layer needed to construct a good garden. For this reason they are nearly always evergreen. Shrubs play a crucial role during the winter months when perennials are cut down and many of the trees and hedges are devoid of their foliage.

Shrubs have long been slaves to gardening fashions. Camellias and azaleas were the first choice during the last century. If a particular type of shrub was a favourite of your parents or grandparents, you might find them old-fashioned and want to avoid planting it in your own garden. The current favourites are *Viburnum* (see page 142) and *Rhaphiolepis* (see page 122). However, some shrubs have endured the vagaries of fashion: *Buxus* (see page 27), *Gardenia* (see page 96) and *Daphne* (see page 93) are examples of these.

There are several shrubs I would love to be able to keep planting but have been forced virtually to abandon because of the heat and drought that has been a sad consequence of climate change. Many of the viburnums and most varieties of camellias and azaleas are just too thirsty to grow in most Australian soils.

Because shrubs are permanent features in the garden, it's important that they are chosen carefully. If you are choosing a shrub for background planting or as a filler plant, choose an evergreen plant with a leaf colour that sets off the rest of the garden. I love shrubs with a dark green leaf because I find they provide an excellent backdrop for the flowering plants. But if you are after a grey foliage garden, try shrubs with sage-green or grey foliage.

Above: *Gardenia augusta* 'Florida'

There are a few tried-and-true combinations of shrubs that I always come back to in my garden designs. They are:

- *Daphne* (see page 93) and *Camellia sasanqua* (see page 88)
- *Murraya paniculata* (see page 52), *Buxus sempervirens* (see page 27) and *Gardenia* (see page 96)
- *Rhaphiolepis* (see page 122) and *Gardenia* (see page 96).

Keep in mind that shrubs tend to look good when used en masse, because their form on their own is not very striking. The general rule is the smaller the shrub the more you need to group together to form a mass. Large shrubs such as viburnums may require only three specimens as a minimum, while gardenias and daphnes need at least six to form a decent group planting.

Tips for shrubs

1. Many shrubs are sensitive to position – keep this in mind as you choose a shrub for a full sun, partial shade or full shade position.
2. Make sure your shrub is not going to grow too large for its position. Nothing looks worse than shrubs that have been over-pruned to fit a small space.
3. Remember that shrubs provide a vital strength and substance to a garden bed, so allow enough width in the garden bed to accommodate them.
4. If you are going to grow a shrub as a feature plant rather than as a background plant en masse, be particularly careful when you're at the nursery to select one with a strong, well-pruned shape.

Abutilon × hybridum
Chinese lantern

Mature height and spread: 2–3 m **Spacing**: 1 plant per 4 m^2 **Plant form/habit**: Irregular **Position**: Full sun to full shade **Leaf hold**: Evergreen **Benefits**: Flowers, foliage **Flowering time**: Mar–Nov **Flower colour**: Multiple **Climate**: Cool temperate, warm temperate, subtropical, tropical **Water requirement**: Drought tolerant

Chinese lantern is not a plant I use often or en masse, but rather as a specimen or focal plant in the garden. The form of the plant allows it to hang rather gracefully over other plants in the bed, thus focusing attention on the bell-shaped flowers that hang from its arching branches. For this reason I would tend to use it either in the corners of garden beds or at accent points, such as either side of an architectural feature. Birds love this plant, and it's one I regularly use when a client requests a plant that will attract birds.

The Chinese lantern will grow best in full sun but does tolerate dry shade. It does not like humidity and will cope well with cold winter conditions. If you position it to the back of the bed it will still fall forward to the front. This fast-growing plant does require some heavy pruning from time to time to maintain foliage density, but very little maintenance aside from that.

A range of colours is available, from warm oranges to creams to white. I prefer the pale yellow varieties; these tone in well with a soft colour scheme. You should limit the number of these plants you use in the garden; I often use just one or a pair.

Buddleja davidii hybrids
Butterfly bush

Mature height and spread: 2–4 m **Spacing**: 1 plant per 4 m^2 **Plant form/habit**: Vase **Position**: Full sun **Leaf hold**: Evergreen **Benefits**: Flowers, foliage, fragrance **Flowering time**: Dec–May **Flower colour**: Multiple **Climate**: Cool temperate, warm temperate, subtropical **Water requirement**: Drought tolerant

I use butterfly bush in a similar fashion to the Chinese lantern (see opposite); that is, as a specimen plant. Birds like it but more importantly it is great for attracting butterflies. I don't use this plant often as it is not the most attractive of shrubs when not in flower, but it is what I call a good 'filler' – it flowers well and fills a deep bed, so serves as a very good background shrub.

Butterfly bush is fast-growing and relatively pest- and disease-free. It is considered weedy in some areas, so check with your local authority before planting and prune off dead flowers to prevent seeding.

It has a long flowering period with large racemes ranging from white to pink to blue, and boasting a most delicious fragrance. Every year it is worthwhile pruning out or cutting back some of the older stems so that new fresh wood can develop and contribute to an ever-welcome floral display. If it does grow out of shape it will respond well to a massive prune, almost to the ground. You will then find it responds with new, fresher foliage and will also flower better.

It is more upright in form than the Chinese lantern, so still plant it at the rear of the garden bed as it will easily grow to 4 metres and its profuse flowering habit will be noticed. It tolerates a lot of heat and requires very little supplementary water. It will tolerate some humidity, but not a lot, and requires a freely draining soil. This plant needs very little maintenance – just some regular pruning to maintain a desired shape.

Camellia sasanqua
Sasanqua camellia

Mature height and spread: 2–4 m **Spacing**: 1 plant per m^2 **Plant form/habit**: Irregular **Position**: Full sun to full shade **Leaf hold**: Evergreen **Benefits**: Flowers, foliage; some cultivars are fragrant **Flowering time**: Mar–June **Flower colour**: Multiple **Climate**: Cold, cool temperate, warm temperate **Water requirement**: Moderate

Like azaleas, camellias have suffered badly due to years of drought, with the leafy inner-city suburbs of Melbourne and Adelaide now almost devoid of them. The *Camellia japonica* species has suffered the most, with new plantings of these in the southern states not taking and older specimens declining badly. The only species I would consider planting now is the *Camellia sasanqua*, which seems to tolerate the heat much better. It is often used as a hedge but I prefer to use it as a good flowering background plant to a deep garden bed or even on its own in a densely planted bed, almost as a very loose hedge.

It boasts a variety of colours, from white and creams to reds and pinks. This camellia flowers prolifically during autumn into winter when there is often not much else out in the garden. My favourite cultivar is 'Setsugekka'. There are some perfumed varieties as well, such as 'Cinnamon Cindy', another favourite.

Sasanquas grow best in milder summer climates in full sun, but I find they do best in partial shade. They will require water in summer, so make sure you have adequate supplies.

This shrub works in gardens that have soils enriched with compost and a regular water supply, though they can get through the dryer times with heavy mulching. Sasanqua camellias have mid-sized, dark, glossy green leaves that form the perfect foil for white-flowered varieties such as 'Setsugekka' with its undulating petals and yellow boss of stamens at the centre of each bloom.

A light clip and feed each year following flowering encourages nice bushy plants.

Caryopteris × clandonensis (C. mongolica × C. incana)
Bluebeard

Mature height: 1 m **Mature spread:** 1 m **Spacing:** 1 plant per 1 m^2 **Plant form/habit:** Rounded **Position:** Full sun **Leaf hold:** Deciduous **Benefits:** Flowers, aromatic foliage **Flowering time:** Late summer to autumn **Flower colour:** Deep blue **Climate:** Cold, cool temperate, warm temperate **Water requirement:** Drought tolerant

I have been experimenting with planting combinations in my herbaceous perennials for some time now and the latest discovery is the *Caryopteris* × *clandonensis*. This small, blue-flowering woody subshrub has been a welcome addition to my flowering palette. I have used it to supplement or replace *Nierembergia* 'Blue Wonder', as this plant can often be difficult to regenerate over the winter period, particularly in cold areas. *Caryopteris* is extremely reliable, flowering for a period of three to four months continuously from late summer well into autumn.

It doesn't require much water, will tolerate extreme heat and grows well on alkaline soils. The blue hue of its flowers is a vibrant addition to any flowering border, and it is attractive to butterflies. I prefer to use this plant in a border and not as a stand-alone plant, as its foliage is sparse and not that attractive. Cut it back to approximately one-third of its foliage size in late autumn to early winter.

Ceratostigma griffithii
Burmese plumbago

Mature height: 500–1000 mm **Mature spread**: 500–1000 mm **Spacing**: 4–6 plants per m^2 **Plant form/habit**: Spreading **Position**: Full sun to partial shade **Leaf hold**: Deciduous **Benefits**: Flowers, foliage **Flowering time**: Feb–May **Flower colour**: Blue **Climate**: Cool temperate, warm temperate **Water requirement**: Drought tolerant

This tough plant is one of the best shrubs for cold winters and hot dry summers. I use it to create a carpet of foliage that masses at about a metre in height with soft green foliage in the summer, turning a magnificent hot red in the autumn. At this time the shrub also becomes smothered in blue flowers. I often plant it under maples as not only does it complement their autumnal colour change, but the two plants co-exist extremely well together.

I recommend cutting the plants to the ground in late winter just before spring growth appears. It will emerge again to its full height within only a month!

I have been growing Burmese plumbago at my country property Stonefields for years with very little water, and it seems to thrive both in full sun and partial shade. It's a great plant to use in beds that are very deep and require a mass planting to fill them, too.

Fortunately this plant seems to be disease- and pest-resistant. However, it will not tolerate wet feet, and poor drainage quickly turns the leaf colour red.

Choisya ternata
Mexican orange blossom

Mature height and spread: 1–2 m **Spacing**: 1 plant per m^2 **Plant form/habit**: Rounded **Position**: Full sun to full shade **Leaf hold**: Evergreen **Benefits**: Flowers, foliage, fragrance **Flowering time**: Sep–Nov **Flower colour**: White **Climate**: Cool temperate, warm temperate, subtropical **Water requirement**: Drought tolerant

This is a plant that has suffered from 'poorer cousin syndrome'. I always think it is rather similar to *Murraya paniculata* (see page 52) but not quite as nice. They both have a mid-green glossy leaf, they both have perfumed flowers (with the *Murraya* being more sweetly and strongly perfumed), and they can both be clipped into a hedge. However, *Choisya* tolerates extreme cold weather better than the *Murraya* and does not grow as big when left as a shrub. It will also grow better when the area is more deeply shaded or there is a lot of competition from larger trees. So it does have its benefits over *Murraya*. However, given ideal conditions for both I would always choose *Murraya* because it is far more fragrant.

Choisya is one of those rare plants that seem to thrive on neglect, providing it is placed in well-drained soil so it can't get wet feet. It makes a great background evergreen shrub for a flowering border and can easily be shaped to whatever height or shape is required. It seems to be relatively pest-free and requires little maintenance.

Plants that are grown with more water or food will create leaves that are more palatable for chewing insects, so keep an eye out for holes in the leaves or try to treat the plants mean so that leaves can harden to discourage pests! Chompers aside, this a good tough plant that will reward with tidy foliage and fragrant flowers in the spring.

Daphne odora
Daphne

Mature height and spread: 1–2 m **Spacing**: 4 plants per m^2 **Plant form/habit**: Irregular **Position**: Partial to full shade **Leaf hold**: Evergreen **Benefits**: Flowers, foliage, fragrance **Flowering time**: July–Sep **Flower colour**: Pale pink **Climate**: Cold, cool temperate, warm temperate **Water requirement**: Moderate

There is no better plant for winter perfume than daphne. Its sweet smell is always a reminder of cold southern winter days. Plant it near your front entrance or under a window and the perfume will be intoxicating. I like planting it either en masse or in a row, creating a very casual low hedge, particularly along small side gardens next to a house.

Be warned: daphne is a very difficult plant to grow, tolerating a narrow range of conditions. For it to thrive it must receive no humidity in the summer, be in either deep shade or very limited filtered morning sun, exist in extremely well-drained soil and be free from much competition from larger tree roots. (Appropriate drainage is the first and foremost criterion to be met when growing daphne.) I find that daphne really only does well in the southern states and when planted in the shade. In Sydney it will grow with difficulty, but given good ventilation it will tolerate a bit more sun. The frustrating thing about growing daphne is that you can have 10 in a row and the middle one will die for no apparent reason.

Plant daphne in rich, well-composted soil and nurture regularly with liquid seaweed and compost. Trim daphne lightly following flowering for maximum young growth.

Daphne × transatlantica 'Eternal Fragrance'

Eternal fragrance

Mature height: 600 mm **Mature spread:** 900 mm **Spacing:** 5 plants per 3 m^2 **Plant form/habit:** Rounded **Position:** Full sun to partial shade **Leaf hold:** Evergreen **Benefits:** Perfumed flowers, year-round tidy foliage **Flowering time:** Mostly in spring, also summer and autumn **Flower colour:** White with pink reverse **Climate:** Cold, cool temperate, warm temperate **Water requirement:** Dry tolerant, but requires watering during extended heat

I have been using the species *Daphne odora* (see page 93) extensively throughout my career; its perfume and time of flowering when little else is out makes it a welcome addition to any garden. However, it had some major drawbacks – a high rate of failure and limited geographical suitability, as it is intolerant to humidity. *Daphne* 'Eternal Fragrance' has been bred for its hardiness and greater range of planting. It will tolerate some humidity, doing much better in Sydney than *Daphne odora*, and has a 90 per cent success rate. It's much smaller than *Daphne odora* and has more prolific but smaller flowers. The perfume is not as strong or as sweet as *Daphne odora*. I use it as a mass planting shrub in semi-shade to full shade situations. It needs good drainage and will quickly fail given wet feet. Tip prune after flowering. It's a great companion to maples, where I use it to carpet the ground under a copse or single maple tree, and is also good as a low hedge or border, or grown in pots.

Escallonia 'Iveyi'
Escallonia

Mature height and spread: 1–2 m **Spacing**: 4 plants per m^2 **Plant form/habit**: Irregular **Position**: Full sun to partial shade **Leaf hold**: Evergreen **Benefits**: Flowers, foliage, fragrance **Flowering time**: Feb–May **Flower colour**: White **Climate**: Cool temperate, warm temperate **Water requirement**: Drought tolerant

This low-care shrub I use only as a mass planting as I feel it's not really good enough to stand alone. I create a large 'wave' in the garden by planting it densely to form one large mass to fill an entire bed. Its dense and very fine foliage lends itself well to this form. It clips extremely well and grows at a moderate rate, which means it does not require excessive pruning.

Escallonia does have a small white flower that is attractive in late summer and into autumn, but it's not abundant. Also, if you shape the shrub as I suggest above, the flower is suppressed by constant clipping.

Escallonia likes full sun or partial shade, a small amount of watering in summer and low humidity. It's not fussy about soil type, but does require good drainage. It provides a foliage screen and is very suitable for coastal locations.

Gardenia augusta 'Florida'
Florida gardenia

Mature height and spread: 500–1000 mm **Spacing**: 1 plant per m^2 **Plant form/habit**: Rounded **Position**: Full sun to partial shade **Leaf hold**: Evergreen **Benefits**: Flowers, foliage, fragrance **Flowering time**: Oct–Apr **Flower colour**: White **Climate**: Warm temperate, subtropical, tropical **Water requirement**: Moderate

This classical plant boasts all the attributes you could ask for: mid-sized, glossy dark-green foliage which sets the background for a display of exquisite succulent white flowers over the summer and autumn. Its perfume can fill an entire garden and is especially noticeable on a warm summer's evening.

The gardenia is one of the hardiest of the shrubs I use. Given the right conditions it thrives, flowering for up to seven or eight months of the year. The perfect planting companion for gardenias, which love the summer and sun, is daphne (see page 93) for the shade and for winter flowering. Gardenias require full, hot blazing sun to flower and thrive. It is a myth that gardenias are shade-dwelling plants. I find they do best on a rooftop in full sun and even behind a glass pool fence where the sun's rays are magnified. It cannot get hot enough for gardenias!

They do well in all states except for Tasmania, as they do not like extreme cold, wind, frost or snow. A protected position is often best and success is achieved in the southern states when planted near the protection and warmth offered by buildings.

They do tolerate cool winters but will not flower during this period.

Good soil preparation is essential for gardenias: cultivate the soil to a depth of just over 600 millimetres, adding lots of rich compost and ensure good drainage. Gardenias hate having wet feet so make sure the drainage is excellent or the leaves will yellow.

There seems to be a lot of confusion around the correct fertiliser for gardenias to maintain leaf colour and good flowering. I find they need a fair amount of nitrogen, and then I use Epsom salts to help them flower.

While gardenias are relatively disease-free, the catch is that they are very tasty to all manner of pests – from aphids to scale, caterpillars and mite. Overhead watering will keep the mites at bay, while monitoring and monthly spraying over summer and autumn is required to keep other pests controlled.

Hydrangea arborescens 'Annabelle'
Snowhill hydrangea

Mature height and spread: 2–4 m **Spacing**: 1 plant per m^2 **Plant form/habit**: Rounded **Position**: Full sun to full shade **Leaf hold**: Deciduous **Benefits**: Flowers, foliage **Flowering time**: Dec–Feb **Flower colour**: Green turning white **Climate**: Cold, cool temperate, warm temperate **Water requirement**: Moisture-loving

This hydrangea is one of my favourites, with its delightful creamy, frothy flower heads that start out green and only turn white at the end of their flowering. They fit perfectly into a totally green garden. I used this hydrangea extensively at my last farm, St Ambrose. Unfortunately like most hydrangeas it does not tolerate the extreme heat of southern summers very well – it will either wilt or burn given hot westerly sun. If planted in dappled shade and given plenty of water it will grow well, however.

This hydrangea will get tougher and more sun-resistant the older it gets and its flowering will become more robust and prolific. Prune it hard in the winter, down to a hard wood branch structure approximately 300–400 millimetres from the ground, then cut back to this same position every year. It is prone to mildew, so it's not wise to plant it in humid regions.

Hydrangea macrophylla
Common hydrangea

Mature height and spread: 1–2 m **Spacing**: 1 plant per 1.5 m^2 **Plant form/habit**: Rounded **Position**: Full sun to full shade **Leaf hold**: Deciduous **Benefits**: Flowers, foliage **Flowering time**: Dec–Feb **Flower colour**: Multiple **Climate**: Cold, cool temperate, warm temperate, subtropical **Water requirement**: Moisture-loving

Hydrangea macrophylla is the hydrangea most people think of when they imagine old-fashioned flowers. It's the hydrangea your grandmother probably grew. Depending on the variety, this species produces either large ball-shaped inflorescences (mopheads) or elegant lace-capped flower heads in blue, pink, red and white.

Flower colour is determined by the genetics of the plant and the soil pH. Acid soils promote blue flowers while alkaline soils produce pink to red flowers. White hydrangeas are not affected by soil pH. I love using this plant en masse: nothing looks better than a walk or massed bed of white or blue hydrangea!

Hydrangea macrophylla is the most prolific flowerer of all the hydrangeas, and it's probably the showiest too. It grows well in humid regions, where it can often be found flourishing in full sun. In the southern states it is now burning or wilting badly during the summer periods, so much so that I have almost stopped using it in these regions. It can only be grown there with a lot of water and dappled shade – even then it tends to wilt in extreme heat, and a drooping hydrangea is most unattractive. Avoid planting in a hot westerly aspect and keep soil moist. Give each plant plenty of room as they can grow to very large proportions, so position one every 1.5 metres.

At the end of the flowering season, you can remove the spent flower stems by cutting them well back to just above a pair of plump buds. This will encourage new wood to grow and new buds to set for flowers in the following season. In areas with harsh winters, it is best to leave the flowers on the plant over the winter to act as frost protection and then, in spring, prune them back moderately, taking care not to remove too many buds as these will produce the flowers for the current season. Always prune with a view to renewal and be sure to remove any dead or tired wood.

Hydrangea paniculata
Panicle hydrangea

Mature height and spread: 2–4 m **Spacing**: 1 plant per 2–3 m^2 **Plant form/habit**: Rounded **Position**: Full sun to full shade **Leaf hold**: Deciduous **Benefits**: Flowers, foliage **Flowering time**: Jan–May **Flower colour**: White **Climate**: Cold, cool temperate, warm temperate **Water requirement**: Moisture-loving

This hydrangea carries large conical panicles held well above its foliage. I first saw it growing like a tree in the Hamptons just outside New York. There you can buy them as small trees (at a very old age, presumably). They are prone to burning in summer in all areas of Australia, whereas in the Hamptons they grow them in full sun! The flowers are not abundant but each flower is quite spectacular: they form a creamy white pyramid shape, almost resembling a large bunch of grapes. They don't flower for long and are not terribly attractive when not in flower, so position them at the back of the bed and don't make them a feature.

This hydrangea grows best in the cooler mountainous regions: the mild weather and often deep rich soil combine to create perfect conditions. Dappled shade is ideal; the plant requires even soil moisture and protection from hot westerly sun and strong winds. It will tolerate some humidity, however. It can be planted as a walk, spaced 2–3 metres apart. I would not plant it en masse but rather use it as a background plant that comes to the fore when in flower.

Flowers are borne on the new season's wood, so in winter prune out the previous year's stems back to one or two buds until you are left with a woody framework.

Hydrangea quercifolia
Oak leaf hydrangea

Mature height and spread: 1–2 m **Spacing**: 1 plant per 750–1000 mm² for a massed-planting effect **Plant form/habit**: Irregular **Position**: Full sun to almost full shade **Leaf hold**: Deciduous **Benefits**: Flowers, foliage **Flowering time**: Jan–Mar **Flower colour**: White **Climate**: Cold, cool temperate, warm temperate **Water requirement**: Moderate

Out of all the hydrangeas, the oak leaf is probably the one that best tolerates the sun and requires a bit less water. For those desperate for hydrangeas in the southern states, this is the one I always recommend. It is also more architectural or sculptural in its appearance, and for this reason suits both a traditional and more contemporary-style garden. It does not flower for as long as *Hydrangea macrophylla* (see page 99), but with its large conical cream flowers it still makes for an interesting show for up to two months in summer.

This hydrangea boasts attractive large textural foliage shaped like oak leaves, as the name implies. I believe these dark green glossy leaves are as interesting as the flowers, so I like to use the plant prominently in the garden bed. The autumnal foliage displays are impressive, with russet reds and rich browns held for a time before leaf drop.

Oak leaf hydrangeas prefer afternoon shade and some morning sun, but also grow in almost full shade – just not deep shade. Prune it the same way as *Hydrangea arborescens* (see page 98); that is, back to the same branch structure every year. These should be approximately 500 millimetres from ground level.

Lavandula angustifolia
English lavender

Mature height and spread: 500–1000 mm **Spacing**: 4 plants per m^2 **Plant form/habit**: Rounded **Position**: Full sun **Leaf hold**: Evergreen **Benefits**: Flowers, foliage, fragrance **Flowering time**: Dec–Feb **Flower colour**: Light purple **Climate**: Cool temperate, warm temperate **Water requirement**: Drought tolerant

Lavender is one of the most useful plants, with its silvery-grey mounds of foliage, producing a purple haze of flowers at the ends of tall stems every summer. The flowers are beautifully and heavily scented, especially when crushed. Not only is lavender extremely tough and drought-resistant but it also crosses over many design styles. It is as much home in a Mediterranean-style garden as it is in an English flower border or en masse in a highly minimalistic contemporary garden.

English lavender, as it is commonly called here, flowers around Christmas with one flush only, while French lavender (see page 108) provides a display almost all year round. English lavender grows easily in the southern regions of Australia (particularly in dry and coastal areas) but unfortunately will not tolerate any humidity. Many of my clients have been determined to grow it north of Sydney, only to be faced with replacing it every summer or simply giving up.

Well-drained soil is paramount: add coarse sand if necessary. English lavender is slightly more tolerant of soil variation than French lavender, however. Make sure your lavender is in full sun, as it will not tolerate any shade. I plant lavender bushes 500 millimetres apart, or four per square metre. These plants are carefree, except you will need to remove flower stems by trimming back into the foliage every year after flowering.

My trick with all lavenders is one I learnt in France. Shape your lavender to form a sphere. Let it flower, where it will fluff out to be a less formal sphere, and then always cut it back to shape. For English lavender this is a once-yearly exercise. If you prune this way your lavender will never become woody, out of shape or prone to mildew, and will flower more prolifically.

Lavandula dentata
French lavender

Mature height and spread: 500–1000 mm **Spacing**: 4 plants per m^2 **Plant form/habit**: Rounded **Position**: Full sun **Leaf hold**: Evergreen **Benefits**: Flowers, foliage, fragrance **Flowering time**: Year-round **Flower colour**: Purple **Climate**: Cool temperate, warm temperate **Water requirement**: Drought tolerant

I use French lavender in a similar fashion to English lavender (see page 104). I always maintain it in a ball shape, as this way it will flower for up to nine or ten months of the year. Allow it to flower for several months and then prune it back – it will come straight back into flower. I often use French over English lavender simply because of its extended flowering season. It is marvellous in a mixed border where it often adds winter flowering, or used en masse or even as a small flowering hedge.

Like its English cousin, French lavender hates humidity and enjoys full hot sun, although it will tolerate a small degree of shade. It is a little more sensitive than the English variety, requiring extremely well-drained soil, and it will benefit from the addition of compost. It is perfectly suited to sandy soils and loves a coastal location. This plant flowers throughout the year but particularly in the winter months. The flowers are more solid than the dainty English ones, and the colour is paler. The key to its success is constant pruning.

Lavandula stoechas 'Avonview'
Spanish lavender, Italian lavender

Mature height and spread: 500–700 mm **Spacing**: 5 plants per m^2 **Plant form/habit**: Rounded **Position**: Full sun **Leaf hold**: Evergreen **Benefits**: Flowers, foliage, fragrance **Flowering time**: Year-round **Flower colour**: Deep purple **Climate**: Cool temperate, warm temperate **Water requirement**: Drought tolerant

Spanish lavender has the showiest flowers and a long flowering period. Deep purple bracts stand up like rabbits' ears at the top of each flower and they are brilliantly coloured, especially in morning or afternoon light, which seems to glow through the purple bracts. Spanish lavender includes many varieties and some are considered weeds, so check with your local authority before planting.

Out of all the lavenders this species has the best chance of growing in a humid environment – some are successful in Brisbane, although they have to be replaced every couple of years. 'Avonview' I use for its neat, compact shape and for its intense purple colour. Its deep, almost azure hue is very eye-catching in the garden and commands attention wherever it's planted.

Flowers are borne all year and this makes it a hard decision as to when to prune back to promote nice bushy growth. The best time I believe is late winter, just prior to the new flush of growth brought about by the warmth of spring.

As with all lavenders, Spanish lavender is pest- and disease-free. Just place them carefully as they attract bees. It is slightly smaller than the English or French plant, so bushes can be positioned more closely together.

Lavandula stoechas is weedy in some areas, so check with your local authority before planting and always grow named varieties.

Magnolia 'White Caviar'
Michelia 'White Caviar'
'White Caviar' magnolia

Mature height: 3–4 m **Mature spread:** 2 m **Spacing:** 1 plant per 3 m^2, 1 m apart if planted as hedge **Plant form/habit:** Upright shrub, can be pruned to hedge **Position:** Full sun to partial shade **Leaf hold:** Evergreen **Benefits:** Perfumed flowers, glossy green foliage **Flowering time:** Spring and summer **Flower colour:** Cream **Climate:** Warm temperate, subtropical, tropical **Water requirement:** Moderate

This is a relatively new variety of *Magnolia*, formerly classified as *Michelia*. It is a hybrid between *Magnolia yunnanensis* and *Magnolia figo*, and I now use it in preference to *Magnolia doltsopa*. It holds its foliage much better than *M. doltsopa*, which tends to droop in leaf straight after flowering, causing the plant to appear as if it is dying. 'White Caviar' retains its leaf form during and after flowering. The leaf is an apple green and doesn't have any of the browning found in some *Magnolias*. The flower is white and has a sweet perfume.

I use it as a background shrub for screening with the added bonus of flowering, rather than as a specimen; I find its form not spectacular enough to feature on its own. It's very reliable and will tolerate a range of conditions, including some humidity, but it does require watering in the drier summers of the southern states of Australia and will not tolerate alkaline soils.

Nerium oleander
Oleander

Mature height and spread: 2–4 m **Spacing**: 1 plant per 2 m^2 **Plant form/habit**: Rounded **Position**: Full sun **Leaf hold**: Evergreen **Benefits**: Flowers, foliage **Flowering time**: Nov–Mar **Flower colour**: Multiple **Climate**: Warm temperate, subtropical, tropical **Water requirement**: Drought tolerant

Oleander is a wonderful evergreen plant, either as a large background shrub for a flower bed or in larger gardens as a filler for entire beds. It's a good informal screening plant. I often position it in a mid to rear tier in a multi-tiered planting layout. One of the reasons for planting it at the back of a bed is that oleander is poisonous to humans and animals if ingested, and contact with the foliage may irritate the skin. It's particularly important to avoid the white sap that it exudes when a branch is torn.

The oleander flowers for almost half the year with good-sized blooms boasting many colours, including pinks, whites, reds, apricot and rose. White is my favourite. The plant grows to form a large shrub and can be clipped into a loose casual hedge or rounded to form large spheres. When used en masse these have a desirable sculptural quality. I find them very handy in large country gardens where water is scarce and garden beds are often very deep, requiring a large shrub as a background to smaller, more delicate shrubs or perennials.

This carefree, low-maintenance shrub tolerates drought, frost, wind, some humidity and a little shade. Its ideal environment is a hot, dry summer and a relatively mild winter – it will do well in coastal locations.

Prune quite heavily as oleander will respond well and often flower more profusely after cutting back. I prune them into a very loose sphere shape, as this helps prevent the shrub dying around the base.

Osmanthus spp.
Osmanthus

Mature height and spread: 1–2 m **Spacing**: 1 plant per 4 m^2 **Plant form/habit**: Irregular **Position**: Full sun to full shade **Leaf hold**: Evergreen **Benefits**: Flowers, foliage, fragrance **Flowering time**: Jun–Aug **Flower colour**: White **Climate**: Cold, cool temperate, warm temperate **Water requirement**: Drought tolerant

Osmanthus is an excellent flowering and fragrant small shrub for any bed. Its foliage reminds me of holly without the prickly spikes. It can be grown as a medium hedge but I find it too uneven and not dense enough. It is relatively slow growing, so don't expect immediate results. The flower is white and is fairly insignificant-looking, but its perfume certainly compensates for this. The foliage is dark green and very vertical in nature and therefore makes for a great background shrub.

There are a few different types of osmanthus I use commonly.

Holly leaf osmanthus (*Osmanthus heterophyllus*) is an attractive specimen with dark, glossy green-and-yellow foliage, all twisted and spined like holly. I use it as a sculptural plant. The shrub is sweetly scented with flowers towards the end of summer. It tolerates all light conditions and is visually textural, making it an interesting hedge with a good depth of colour.

Fragrant osmanthus (*Osmanthus fragrans*) emits the best fragrance of all the osmanthus – a fruity, sugary scent that is quite unique. It has large, dark green leathery leaves and is more suited to background planting because its large internodes create long woody stems, making it look a little untidy. Its form and foliage are rather boring, in my view, so place this one in a spot that does not feature the plant but still allows the perfume to fill the air. Note that it grows larger than *O. heterophyllus* and *O. × fortunei*, so it belongs at the rear of a very deep bed. This is a good plant to keep in mind for its ability to take all degrees of light as well as its unforgettable fragrance.

Fortune's osmanthus (*Osmanthus × fortunei*) is a hybrid between *O. heterophyllus* and *O. fragrans*. This is probably the most robust of all the osmanthus and that is saying something, because this plant is synonymous with toughness. Mid-sized thick, leathery leaves are a dull dark green and semi spined, so the leaf resembles holly. It needs to be trimmed regularly to encourage bushy growth and good stem coverage. This tough hedge tolerates dry periods and all levels of light – and has the added bonus of scented flowers in late summer.

Philadelphus coronarius
Mock orange

Mature height and spread: 1–2 m **Spacing**: 1 plant per 3 m^2 **Plant form/habit**: Vase **Position**: Full sun to partial shade **Leaf hold**: Deciduous **Benefits**: Flowers, foliage, fragrance **Flowering time**: Nov–Jan **Flower colour**: White **Climate**: Cold, cool temperate, warm temperate **Water requirement**: Moderate

Mock orange is a tough plant, grown for its rewarding scented white flowers that are produced on large upright and arching canes during late spring and early summer. This plant reminds me of my childhood in gardens full of summer perfumes and mostly white colour schemes. It is a very old-fashioned plant but has endured mostly due to its sweet, citrus-like summer aroma. Its scent can permeate the whole garden.

Make sure it is placed so when in blossom the flowers can be reached, for the first thing people will do when walking into a garden with mock orange is grab the nearest branch and bring the flower closer to smell its perfume.

It has not fared that well during the drought and heat of the past years, but will do well in cooler climates and mountain areas where it thrives with dogwoods and viburnums. If planting in southern capitals choose a sheltered, lightly shaded area away from the afternoon sun. It grows quite tall, so plant to the rear of the garden bed and use as a specimen maybe in the corners or in highly visual points in the garden.

Philadelphus is a tolerant genus, coping with heat, cold and short dry periods. It is a low-care plant except that it needs to be pruned annually to promote new growth and greater flowering. Pruning should take place after flowering with one-third of the old stems cut down to ground level. Shorten the stems remaining by about one-third by pruning back to where new stems have developed. Retaining the old stems is important because these will generate the flowers for the following year. Feed plants following pruning, so that new roots and shoots can develop with vigour.

Pittosporum tobira
Japanese pittosporum

Mature height and spread: 2–4 m **Spacing**: 2 plants per m^2 **Plant form/habit**: Rounded **Position**: Full sun to full shade **Leaf hold**: Evergreen **Benefits**: Flowers, foliage, fragrance **Flowering time**: Jan–Mar **Flower colour**: Cream **Climate**: Cool temperate, warm temperate, subtropical **Water requirement**: Drought tolerant

This tidy evergreen pittosporum boasts lovely cream-coloured, scented flowers in the summer. I first saw it growing in the south of France, where it is very popular for its sweet perfume. I thought it so good that I used it extensively in a garden I created in St Tropez. It has a dark, shiny green glossy leaf that contrasts well with the often matt greys of other Mediterranean plants. I use it en masse with great success, often to fill an entire bed or to complement the mass planting of lavender (see pages 104–9) or *Westringia* (see page 144). A great compact form is also available: 'Miss Muffett'.

Japanese pittosporum is extremely tough and tolerates a hot dry summer as well as poor or sandy soil conditions. It loves full sun but will tolerate some limited shade and a little humidity. It forms a neat button shape without any pruning, thus lending itself to a garden full of sphere shapes, such as with clipped lavender and *Westringia* balls. The plant is better suited to harsh or exposed conditions such as coastal areas, because in a protected garden sanctuary new growth is commonly attacked by pests and requires regular spraying to avoid disappointment.

Plumbago auriculata 'Royal Cape'
Plumbago

Mature height and spread: 1–2 m **Spacing**: 1 plant per m^2 **Plant form/habit**: Spreading **Position**: Full sun to partial shade **Leaf hold**: Evergreen **Benefits**: Flowers, foliage **Flowering time**: Mar–May **Flower colour**: Blue **Climate**: Warm temperate, subtropical, tropical **Water requirement**: Drought tolerant

I have often used plumbago en masse as a single species per bed in a diverse range of situations. From humid to extremely dry conditions, this plant manages to thrive and produce an amazing display of bright, pale and deep blue flowers for the entire season of summer to autumn. 'Royal Cape' is my preferred cultivar with a good depth of blue in the flowers. Use it en masse as I do and don't mix it with other plants unless in combination with another plant en masse, for example, lavender (see pages 104–9) or *Nierembergia* (see page 244). It forms a natural flowing shape and requires just a little shaping. It will also spill if placed adjacent to a retaining wall.

This plant is altogether tough. It is pest- and disease-free and takes most ranges of light, though it will flower less in the shade. It loves full sun and a very hot summer, so plant it out in the open and don't be afraid to subject it to all the harsh elements. Once it gets its roots well established, it also takes very dry conditions. This plant is suitable for the coast but probably not recommended for extremely cold locations. I have used it in Queensland all the way down to southern Victoria and it has thrived in all these environments.

Plumbago requires cutting almost to the ground every four to five years, as this will prevent it from becoming woody. Left completely untrimmed, it can get untidy and unruly.

Below: *Plumbago auricalata* 'Royal Cape', *Ligustrum vulgari*

Plumeria acutifolia
Frangipani

Mature height and spread: 2–4 m **Spacing**: Plant 3 m from other trees or large shrubs **Plant form/habit**: Rounded **Position**: Full sun to partial shade **Leaf hold**: Deciduous **Benefits**: Flowers, foliage, fragrance **Flowering time**: Jan–April **Flower colour**: Multiple **Climate**: Warm temperate, subtropical, tropical **Water requirement**: Moderate

A symbol of the tropics, this plant is grown for its deliciously scented flowers. Its flower colour ranges from white and yellow to sunset hues and even red. The foliage is bold and textural and the shape of the plant is usually a well-defined dome. In the dormant period the leaves fall to reveal a candelabra of thick, fleshy stems which are quite sculptural in their own right. I like to use it as a specimen or even as a main focal point in a highly contemporary garden design. I will often place it so it is the centrepiece of a large courtyard or small garden. I use it rather like a small tree.

The frangipani prefers a semi-tropical environment, although it will grow well in the inner suburbs of Melbourne, Sydney, Adelaide and Perth. It hates cold weather and frost in particular. In the southern states it is reasonably slow to grow, while in the tropics it is very fast. Give it plenty of water but make sure the drainage is good. It seems to do well in sandy soils and also tolerates some salt and wind. While preferring full sun, it will grow in semi-shade too.

Rhaphiolepis spp.
Rhaphiolepis

Mature height and spread: 500–1000 mm **Spacing**: 2 plants per m^2 **Plant form/habit**: Rounded **Position**: Full sun to partial shade **Leaf hold**: Evergreen **Benefits**: Flowers, foliage **Flowering time**: Oct–Mar **Flower colour**: White, pale pink **Climate**: Cool temperate, warm temperate, subtropical **Water requirement**: Drought tolerant

I have only recently discovered this hardy plant, while searching for tougher, less water-needy species during the drought. It is extremely tough, requiring little attention, and more importantly very little supplementary water – although it welcomes moisture to the roots. It provides small white or pale pink flowers in the spring and summer and good dark green foliage.

In addition to being drought-proof and pest-resistant, this plant is frost hardy and will tolerate humidity. It requires a lot of sun, however, and will only abide a small degree of shade. Make sure the planting area is well drained, and while it will grow in poor soils, you'll see it thrive in better soil. It is weedy in many areas.

I like to grow *Rhaphiolepis* 'Snow Maiden' en masse, creating a cloud-like effect in the garden. To achieve this, simply prune it once a year into very informal spheres and allow them to then knit together, forming a cloud of dark green foliage. I tend not to use it as an individual specimen as I find it far more attractive in groupings.

Rhaphiolepis 'Oriental Pearl' (see page 122) is another white-flowering form I use frequently in my planting schemes. It is very similar to 'Snow Maiden' in its growing requirements and appearance. I tend to use this variety more as I find it slightly tighter in form and more profuse in its flowering. Great in dappled light under trees, the plant is smothered in white flowers in the summer. Again, this one boasts a tight, naturally domed habit and needs virtually no pruning to maintain this form. New growth begins life as a dull pale green then hardens off to the dark resilient green typical of this species.

Rhaphiolepis × indica 'Oriental Pearl' PBR
Indian hawthorn

Mature height: 1 m **Mature spread:** 1 m **Spacing:** 1 plant per m^2, or 700 mm apart if planted as border **Plant form/habit:** Rounded **Position:** Full sun to full shade **Leaf hold:** Evergreen **Benefits:** Flowers, compact shrub form, dark glossy green foliage **Flowering time:** Early autumn to late spring **Flower colour:** White flowers with pink stamens, gives blush appearance **Climate:** Cool temperate, warm temperate, subtropical **Water requirement:** Drought tolerant

This is the best variety of *Rhaphiolepis* for hardiness and prolific flowering. It will tolerate and even thrive with neglect, loving full sun, and doesn't suffer from extreme heat or frost. It will grow in limited shade but suffers in deep shade. I have seen it grow with tough competition from major trees but it loves growing as a mass planting on its own. I use it as a mass planting small shrub, allowing it to grow as a wave of meandering differing heights, moulded together as one mass. It has a compact habit and rounded, dark glossy green foliage, and is good as a low hedge or to underplant trees. It's sculptural as simple greenery but when it flowers it becomes a mass of the palest pink (tending to white) flowers that last for over a month. It will tolerate coastal conditions, is almost thornless and works well planted beside pools.

Rhododendron indica 'Alba Magnifica'
White evergreen azalea

Mature height and spread: 500–1000 mm **Spacing**: 2–4 plants per m^2 **Plant form/habit**: Rounded **Position**: Full sun to full shade **Leaf hold**: Evergreen **Benefits**: Flowers, foliage **Flowering time**: April–Dec **Flower colour**: White **Climate**: Cold, cool temperate, warm temperate **Water requirement**: Moisture-loving

Azaleas, with their large white flowers from late autumn to early summer, present an uplifting display and have long been a favourite of Australian gardeners in nearly all capital cities and country areas. They tolerate and even seem to thrive in the humid conditions of Brisbane and Sydney, and once did equally well in the colder and drier southern areas. Climate change and the proliferation of the azalea lace bug have had a drastic effect on them, though; so much so that I no longer use them much in the southern states – only the 'Alba Magnifica' variety seems to cope.

'Alba Magnifica' is the toughest of all the azaleas and tolerates heat the best. It is a great plant when used en masse, and this is exactly how I use it: sometimes hundreds of them in the one garden! When used this way they all tend to mould together, and when this occurs it's important not to cut them back as a straight hedge but to prune them so they look as though a wave is passing through the mass of foliage.

All azaleas are prone to the azalea lace bug. This bug starts by laying its eggs on the underside of the leaf: look out for this, as once hatched they quickly eat their way through the leaves, leaving them with a transparent silvered appearance. A regular spraying program over the summer months is a must to ensure show-stopping plants.

Rosa Heritage TM Rosa 'AUSblush'
David Austin English rose

Mature height: 1.2 m **Mature spread:** 1.2 m **Spacing:** 5 plants per 6 m^2 **Plant form/habit:** Rounded **Position:** Full sun **Leaf hold:** Deciduous **Benefits:** Repeat flowers, strongly fragrant **Flowering time:** Oct–May **Flower colour:** Pale pink **Climate:** Cold, cool temperate, warm temperate **Water requirement:** Dry tolerant, but water during extended dry periods

Heritage is the rose I always use when I need a pale pink, recurrently flowering shrub rose. It's very reliable and is spectacular when in flower, but also looks good as a shrub without flowers. I prefer my roses to look full and dense in leaf as well as in flower. The flowers are carried well on the bush and tend not to droop.

The flower itself is perfectly formed and cup-shaped. It also blends well into a flower border and does not rely on mass planting for effect. It's a vigorous shrub rose with a compact habit and relatively few thorns. Deadhead after each flush of blooms. I would only use this rose in the drier states, as it doesn't tolerate any humidity, but it does tolerate heat.

Rosa Iceberg TM Rosa 'KORbin'
Iceberg rose

Mature height: 1.2 m **Mature spread:** 1.2 m **Spacing:** 5 plants per 6 m^2 **Plant form/habit:** Rounded **Position:** Full sun **Leaf hold:** Deciduous **Benefits:** Flowers, glossy disease-resistant foliage **Flowering time:** Oct–May **Flower colour:** White **Climate:** Cold, cool temperate, warm temperate **Water requirement:** Dry tolerant, but water during extended dry periods

This rose was extremely popular in the 1980s with the rise of white-themed gardens. Its popularity was well deserved; it's extremely vigorous and hardy, and blooms prolifically with mildly scented, pure white double blooms. I have recently started using it again due to its reliability and dramatic effect when used en masse in blocks of planting; it will appear as a large cloud of white with little foliage to be seen. Deadhead it regularly and it will continue to flower for up to nine months of the year. Iceberg is also the best choice for a white-flowering rose in humid conditions, such as from Sydney to Brisbane, and it works well as a border planting.

Rosa rugosa var. alba
White rugosa rose

Mature height and spread: 1–2 m **Spacing**: 1 plant per 1.2–1.5 m^2 **Plant form/habit**: Rounded **Position**: Full sun to partial shade **Leaf hold**: Deciduous **Benefits**: Flowers, foliage, fragrance, fruit **Flowering time**: Oct–Jan **Flower colour**: White **Climate**: Cool temperate, warm temperate **Water requirement**: Drought tolerant

Most people generally associate roses with pretty flowery gardens, but I find rugosa roses robust and architectural. They are vigorous, and I like using them in a mass planting or even a hedge. Dark, textural, leathery leaves are a perfect backdrop for clear, large, simple white flowers. I first discovered the benefits of rugosa roses when I was working in the Hamptons: there I saw them used en masse to great effect and, more importantly, I saw them grown in sand dunes. If a plant can survive in sand, it has to be tough! I have since used it to create large swathes of white in large country gardens or to fill entire beds in larger city gardens. I tend to use the white varieties for this purpose and in particular *Rosa rugosa* var. *alba*.

These plants are extremely tough and resilient to both pests and disease; they survive with virtually no water and very little maintenance. They flower throughout the entire rose season and then further reward you with red rosehips (the fleshy fruit that is used for herbal tea) in the autumn. They can be pruned to form one large mass of foliage when planted together unlike other roses, which need to be separated for ventilation. Like all roses, they hate humidity and prefer it hot and dry, so only grow them in the southern states, or in areas of low humidity. They require full sun but will tolerate a small amount of shade.

Trim them lightly to maintain the height and shape you require, not hard like other roses. This is because flowers are borne on the fresh laterals generated from the older wood of the previous year, therefore if you prune this rose back hard you could be removing its potential to flower.

Rosmarinus officinalis (syn. Salvia rosemarinum)

Common rosemary

Mature height and spread: 1–2 m **Spacing**: 1 plant per m^2 **Plant form/habit**: Upright **Position**: Full sun **Leaf hold**: Evergreen **Benefits**: Flowers, foliage, fragrance, culinary herb **Flowering time**: Jan–May **Flower colour**: Pale blue **Climate**: Cool temperate, warm temperate **Water requirement**: Drought tolerant

I love all Mediterranean plants and use them extensively: wisteria, lavender and, in particular, rosemary. We can use rosemary to great effect in certain parts of Australia. Like lavender, it enjoys hot, dry summers with no humidity or cold, wet winters, although the latter is not so essential. It will only do well in the southern states that experience very little humidity, although I have grown it with limited success in Sydney, where excellent ventilation is necessary. Naturally, it is suitable for planting in coastal areas. It flowers with a mass of pale blue small flowers, often in summer and autumn.

The common rosemary form is neither prostrate nor bolt upright, but somewhere in between. While *R. officinalis* is the species most people grow as a herb, I often use it as a hedge, particularly in a herb garden. It can also be employed as a background foil for a more interesting foreground planting. In addition I have grown it at the base of buildings where I clip it into a soft sphere shape. If grown as an individual plant, clip it into a very informal sphere as this helps prevent it from going woody and provides you with a softer, more integrated shape.

It must have good drainage and requires very little in the way of watering. Use this species for cooking, as the other varieties tend to be too pungent.

Rosmarinus 'Tuscan Blue' (syn. Salvia 'Tuscan Blue')
Upright rosemary

Mature height and spread: 1–2 m **Spacing**: 2 plants per m^2 **Plant form/habit**: Upright **Position**: Full sun **Leaf hold**: Evergreen **Benefits**: Flowers, foliage, fragrance **Flowering time**: Jan–May **Flower colour**: Blue **Climate**: Cool temperate, warm temperate **Water requirement**: Drought tolerant

'Tuscan Blue' rosemary flowers more profusely than *Rosmarinus officinalis* and tends to be more vertical and tight in its growth habit. For this reason I use it at the back of smaller beds or as an unclipped informal hedge. Its verticality provides a very stylised, sculptural quality to a contemporary garden.

Enjoying well-drained soil, it tends to flower up the entire stem with aromatic, pretty, deep blue flowers. Happily, it is very low maintenance but it is advisable to prune annually to renew growth.

Do not cook with this variety as the flavour is too pungent.

Santolina chamaecyparissus
Cotton lavender

Mature height and spread: 300–500 mm **Spacing**: 4–6 plants per m^2 or 500 mm apart **Plant form/habit**: Rounded **Position**: Full sun **Leaf hold**: Evergreen **Benefits**: Flowers, foliage **Flowering time**: Dec–Feb **Flower colour**: Yellow **Climate**: Cold, cool temperate, warm temperate **Water requirement**: Drought tolerant

Santolina is a Mediterranean-style plant, so it is drought-hardy and loves a hot dry summer. I often use it with English lavender (see page 104) in particular, as it is of a similar height and shape. It has a yellow flower, which can be attractive in a border, but I prefer to trim it to a sphere and use it as a relief to lavender; for example, in a parterre with half the beds filled with lavender and the other half with santolina.

Like lavender, it hates humidity and wet feet, and requires very little water. It tends to have a limited life, though, and needs replacing every five years. It can also be a bit sensitive – one plant in 10 usually curls up its feet and dies for no apparent reason.

If grown in a flower border, plant several shrubs together to increase the effect, as it tends to be a bit boring on its own. It should be clipped annually around early spring or following flowering to keep the foliage young and tight. Plant only in full sun and make sure the drainage is excellent.

Spiraea cantoniensis
Reeves's spiraea

Mature height and spread: 1–2 m **Spacing**: 1 plant per 2 m^2 **Plant form/habit**: Vase **Position**: Full sun to partial shade **Leaf hold**: Deciduous/semi-deciduous **Benefits**: Flowers, foliage **Flowering time**: Oct–Dec **Flower colour**: White **Climate**: Cool temperate, warm temperate **Water requirement**: Moderate

Some plants just seem to fit together, creating a wonderful array reminiscent of large English flower gardens and therefore early Australian attempts to create this look. Viburnums, dogwoods, rhododendrons, azaleas and spiraeas are the main players in such a scheme. Like most English-style flowering plants, they require a lot of water and are not tolerating the rise in our summer temperatures well. Spiraea is, however, one member of the group that seems to survive these tougher summers.

This quiet background shrub turns on a remarkable display during late spring and early summer when it is smothered in blooms. Masses of umbelliferous flowers are borne on the arching canes that will make you stand back and admire them. For me, this really heralds the beginning of the season.

The medium-sized spiraea shrub grows into a circular shape, thus suiting a position in the middle of a large flower bed. I use it mainly as a single specimen in a flower border or en masse as a background plant in a narrower bed. Spiraea will tolerate semi-shade as well as full sun and will grow in limited humidity. It has the bonus of being semi-deciduous in warmer climates, for a more effective screen. It will need some water during the summer months.

Spiraea requires little maintenance, but a good renewal should take place every two to three years. Prune older stems right to the ground, leaving the opportunity for new wood to emerge for more vigour and flowering. The best time to prune is after the flowers have finished. It is pest- and disease-resistant.

Viburnum odoratissimum
Evergreen viburnum

Mature height and spread: 2–6 m **Spacing**: 1 plant per 1.5–2 m^2 **Plant form/habit**: Rounded **Position**: Full sun to partial shade **Leaf hold**: Evergreen **Benefits**: Flowers, foliage, fragrance **Flowering time**: Nov–Dec **Flower colour**: White **Climate**: Cool temperate, warm temperate, subtropical **Water requirement**: Moderate

There are three or more forms available of this large-leafed evergreen viburnum, the most common being the matt leaf variety or the two shiny leaf forms. 'Awabuki' has thick, durable shiny leaves and 'Emerald Lustre' has even larger leaves which are also glossy and desirable for their visual texture.

'Emerald Lustre' is evergreen, and unlike other viburnums, is grown more for its foliage than its flowers. The flowers are white and in a spray but not profuse. With its relatively large and very dense foliage it makes the plant suitable for hedges and screens and as a simple green background plant to a deep flower bed.

I also favour this viburnum as it is far hardier than the other varieties, faring well during the heat of summer and not requiring much supplementary watering. This plant can be positioned in full sun to part shade; however, it is susceptible to red spider mite in warm, dry summers, such as those experienced in Melbourne, so it fares better in more humid gardens from New South Wales to southern Queensland.

It grows to a height of around 2–6 metres and I would not recommend growing it any lower than this. One of its main attributes is that it trims easily, so keep it tidy by pruning to a large ball shape or trim regularly to maintain a hedge appearance.

Viburnum opulus
Snowball viburnum

Mature height and spread: 2–4 m **Spacing**: 1 plant per 3 m^2 **Plant form/habit**: Irregular **Position**: Full sun to partial shade **Leaf hold**: Deciduous **Benefits**: Flowers, foliage, fruit (inedible) **Flowering time**: Nov–Dec **Flower colour**: White **Climate**: Cold, cool temperate, warm temperate **Water requirement**: Moderate

Below: *Viburnum opulus* 'Roseum' (sterile)

Viburnum opulus appreciates the same growing conditions as *V. plicatum* (see page 142), needing a sheltered, coolish semi-shaded position with low humidity and plenty of summer irrigation. I often plant these together in the same garden as they complement each other, growing to approximately the same height and spread. They flower at slightly different times, which increases the flowering period of the garden. *V. opulus* has large, showy white flowers presented as a round snowball of petals. It flowers prolifically. Similar to *V. plicatum*, the shrub is smothered in them. It also has the added benefit of bright red translucent berries in the autumn, which can be carried through into the winter months for ornamental value.

A favourite cultivar of mine is *V. opulus* 'Roseum', which has extremely showy pom-pom-like flowers. 'Roseum' is a sterile variety, however, so it does not bear fruit.

A more robust viburnum than *V. plicatum*, *V. opulus* can tolerate greater heat and dryness and even cope with wind. This is because the leaves are more durable, being thick and leathery. Flowers occur in late spring to early summer. Like *V. plicatum*, this species tolerates extreme cold and no pruning is required.

Below: Fruit of *Viburnum opulus*

Below: *Viburnum opulus* 'Roseum' (sterile)

Viburnum plicatum forma tomentosum
Viburnum

Mature height and spread: 2–4 m **Spacing**: 1 plant per 2–3 m^2 **Plant form/habit**: Spreading **Position**: Full sun to partial shade **Leaf hold**: Deciduous **Benefits**: Flowers, foliage, fruit (inedible) **Flowering time**: Sep–Nov **Flower colour**: White **Climate**: Cold, cool temperate, warm temperate **Water requirement**: Moisture-loving

In spring, when the flowers of *Viburnum plicatum* sit upright along the length of its branches, the show created is hard to match. The large saucer-shaped flowers are vibrant white – so white they appear to glow in the moonlight. They flower in early spring to accompany the spiraeas (see page 138) and dogwoods (see page 278), creating a perfect flowering spring garden. Both *V. plicatum* 'Mariesii' and *V. plicatum* 'Lanarth' boast even larger, showier sterile flowers.

Also known as the Japanese snowball bush, this viburnum is deciduous but has an interesting lateral branch structure that is very sculptural in the winter. I use it as a specimen shrub, planting it in a prominent position in the garden. If placing it in a large garden bed, position it at the back where it can provide height and substance.

This classy plant is better in cooler, southern regions as it doesn't favour high humidity but tolerates extreme cold. It will burn in extreme heat. Some afternoon shade is advisable, as is protection from wind exposure. Viburnum does need moderate watering during the summer months. It can be left unpruned.

Westringia fruticosa
Native rosemary

Mature height and spread: 1–2 m **Spacing**: 1 plant per lineal 600 mm or 2 plants per m^2 for a massed-planting effect **Plant form/habit**: Rounded **Position**: Full sun to partial shade **Leaf hold**: Evergreen **Benefits**: Flowers, foliage **Flowering time**: Sep–Nov **Flower colour**: White **Climate**: Cold, cool temperate, warm temperate **Water requirement**: Drought tolerant

Westringia is one of our wonderful native plants that fits well into a more exotic planting scheme. Its sage-green foliage colour suits a Mediterranean planting scheme, complementing lavenders and rosemary. It exhibits fine dark green to grey leaves with small white flowers throughout the year. I use it extensively in coastal gardens, where it thrives in sandy soils and salt wind; if planted inland, however, it will require a well-drained soil. Not known for its performance in humid locations, it should be adaptable for alpine regions of Australia.

I like to keep *Westringia* trimmed to a loose sphere shape, or it can simply be left to do its own thing. I often use it en masse with many plants, forming differing-sized spheres. It is extremely drought resistant and will tolerate full sun and some shade, making it a very versatile plant. It also does well in richer soils and will even tolerate cold winters.

Westringia is sometimes sensitive to fertilisers, so avoid manures or fertilisers high in phosphorus around this plant, otherwise it may overfeed to the point of toxicity. This is due to its highly adapted root system, which is able to extract even very small amounts of nutrients from the soil.

Below and right: *Westringia* 'Wynyabbie Gem'

Opposite: *Westringia fruticosa*

Groundcovers

Previous page: *Hedera canariensis*, *Ficus microcarpa* var. *hillii*
Above: *Ajuga reptans*

Whenever I look at a successful garden – one that really captivates me – I ask myself why it works so well. More often than not the reason is because of the depth of its planting. Layers of planting complement one another and build up to form a mature, successful overall scheme. And one of the most important layers is the groundcover.

Groundcovers often provide the foreground for an amazing border, or add further depth to higher planting by weaving in and out like a thread in a beautiful garment. These plants can also be stunning on their own, forming a mass plane of greenery or colour in situations that require little or no height in the planting. If the hedges are the walls of the room then the groundcovers form the vitally important floor covering.

I often use groundcovers to help highlight other features in the garden, such as a tree or an ornament. A highly sculptural or flowering tree is greatly enhanced when viewed rising from a bed of well-formed groundcover. Similarly a modern piece of sculpture seems more dramatic when emerging from a bed of low greenery. This pillow effect, with groundcovers billowing towards a focal point, or blanketing the ground beneath a taller structure, is one of my favourite techniques and one I use often in contemporary gardens.

Groundcovers are often the answer when there is a steep or inaccessible slope and where grass would prove difficult or impossible to maintain. Groundcovers stabilise the soil, stopping soil slippage during periods of rain. They also often reward with wonderful texture and floral displays. I have used banks of ivy (see pages 158 and 184) and star jasmine (see page 200, in the Climbers section – it is both a groundcover and climber) to great success in these situations. The jasmine is particularly rewarding as it flowers prolifically during the summer months, providing you with masses of tiny white flowers and an intoxicating perfume.

Groundcovers not only provide an important visual element by adding depth to the planting scheme, they also reduce the maintenance required for the garden overall. By covering the soil densely with their closely hugging foliage they help suppress weed growth; they also lower the ground temperature, which helps to reduce water loss through evaporation. Groundcover plants may require a certain amount of maintenance while they are being established, but once they are, they prevent weed growth and require very little further maintenance.

In an era of water restrictions and the desire to use fewer pesticides in our gardens, lawns are sometimes not the best choice available. While nothing can beat lawn

Above: *Thymus serpyllum* var. *albus*

for areas of high traffic and where children play, in areas that don't need to tolerate a lot of human traffic and recreation, groundcovers offer an excellent alternative.

Groundcovers are relatively indestructible. They use far less water than lawn and of course don't require mowing. In my opinion they provide an excellent visual replacement for lawn, with all the extra environmental benefits. Some groundcovers, such as creeping white thyme (see page 164), will tolerate being trodden on occasionally, but keep in mind that if this happens regularly, it will start to wear out in patches.

Gardens dominated by groundcovers can be elegantly simple and are ideally suited to owners with destructive dogs or where extremely low maintenance is required. But keep in mind that a groundcover-dominated garden will need some height given to it; this can be provided by adding trees that also require very little maintenance.

One of the main benefits of groundcovers is that they spread rapidly, but this can also be a drawback. Most groundcovers multiply by layering, allowing them to spread laterally at a fast pace through the garden. For this reason they are often best kept in beds where they are the only plant or complement something much larger, such as a tree or large shrub. Make sure the bed is well edged (steel edging being the best choice) so you can easily delineate the boundary. (You can read more about this in my previous book, *Paul Bangay's Garden Design Handbook*.) The best way to control the plant is by mechanical means, with brush-cutters or shears.

And one final warning: in some environments, groundcovers can become weeds. So always check with your local authority to ensure your chosen plant is not on their particular weed lists.

Tips for groundcovers

1. Choose a groundcover that has the right growth habit for the area. You'll want a rampant spreading groundcover for a large area and a slow-growing one for a smaller spot.
2. Make sure your groundcover and the plants it is underplanting are compatible. The groundcover should not dominate the plants it is growing beneath. Keep the groundcover trimmed away from the bases of any shrubs or trees growing above it.
3. Groundcovers are often vulnerable to frost due to their horizontal growth habit. So if you get some frost in your area, choose frost-hardy groundcover varieties.
4. Don't forget that some climbers can be equally at home as groundcovers. Consider star jasmine (see page 200) and *Hardenbergia violacea* (see page 183).

Ajuga reptans
Carpet bugle, ajuga

Mature height: Up to 300 mm **Mature spread**: 500–1000 mm **Spacing**: 9 plants per m^2 **Plant form/habit**: Groundcover **Position**: Partial shade to full shade **Leaf hold**: Evergreen **Benefits**: Flowers, foliage **Flowering time**: Oct–Nov **Flower colour**: Blue, white **Climate**: Cool temperate, warm temperate **Water requirement**: Moisture-loving

Ajuga reptans is a great groundcover for shaded areas where it will grow relatively flat and spread between stepping stones, which prevents people from tripping as they walk from paver to paver. It has small spikes of royal blue flowers that rise approximately 100 millimetres above the foliage. Being a woodland plant, it loves being in damp shaded areas, creeping among the base of trees and shrubs. It does not like too much sun, preferring the filtered light of a woodland setting. Despite this, it has tolerated the recent drought well.

The white-flowering *Ajuga reptans* 'Alba' likes the same conditions as the blue. It is almost exactly the same in form and leaf – the only difference is the flower colour. I do prefer the blue, though, as it is more brilliant than the white, and this is often necessary in the darkness of a woodland. If you are creating a pale colour scheme or an all-white garden, however, 'Alba' is a good choice either as a groundcover or low front border plant. 'Jungle Beauty' is a larger leafed form, with larger flowers also. If a coarser look is required, then 'Jungle Beauty' is the answer.

Below: *Ajuga reptans*, *Euphorbia* sp.
Opposite: *Ajuga reptans* 'Jungle Beauty'

Convolvulus sabatius
Moroccan glory vine

Mature height: 300 mm **Mature spread:** 2 m **Spacing**: 4 plants per m^2 **Plant form/habit**: Trailing **Position**: Full sun to partial shade **Leaf hold**: Evergreen **Benefits:** Flowers, foliage **Flowering time**: Sep–Mar **Flower colour**: Blue **Climate**: Cool temperate, warm temperate, subtropical
Water requirement: Drought tolerant

This must be one of the most invasive and vigorous of groundcovers. For this reason, I only use convolvulus when it can be easily contained; that is, restricted to its own dedicated bed. This plant will reward you with masses of bright blue flowers for seven to eight months of the year. This block style of planting is the only way to use this vine – if you plant it among other plants, it will take over.

Convolvulus will tolerate a wide range of conditions, but does burn back to almost no foliage in hard frost. If this happens, don't panic as it will return in the spring with much vigour.

Its grey foliage is well suited to Mediterranean-style gardens and its blue flower is abundant for most of the year. I find it grows in nearly all zones except those with high humidity. It requires little water, and will not tolerate waterlogging.

Leaders can be trimmed back so they do not become unruly. It is pest-free, disease-free and carefree – what more could you want?

Opposite: *Convolvulus sabatius, Tilia, Lavandula dentata*

Dichondra argentea 'Silver Falls'
Silver falls

Mature height: 200 mm **Mature spread**: 2 m **Spacing**: 300 mm off-centre **Plant form/habit**: Trailing **Position**: Full sun to light shade **Leaf hold**: Evergreen **Benefit**: Foliage **Climate**: Cool to tropical **Water requirement**: Moderate

This cascading silver foliage plant can be used for either falling over the edge of pots as a companion to a larger plant such as olives, citrus or box, or as a groundcover to the front of retaining walls. Its soft grey foliage lightens a planting scheme but its true beauty is the length of the hanging foliage, which can often be up to 1 m in length. It is very drought hardy and loves well-drained soil.

Hedera canariensis
Canary Island ivy

Mature height: 300–400 mm **Mature spread:** 4 m x 4 m **Spacing:** 5 plants per 4 m^2 **Plant form/habit:** Trailing **Position:** Part shade to full shade **Leaf hold:** Evergreen **Benefits:** Foliage **Climate:** Cold, cool temperate, warm temperate **Water requirement:** Drought tolerant

This is the large, glossy-leaved form of ivy and the only one I use as a groundcover. The deep dark green colour to the leaf makes a handsome carpet, especially in dark, shady areas where it will lighten the space. It's very drought-hardy but will not tolerate frost or extreme heat; both cause it to burn, and it grows better in dappled shade to shade. It will tolerate dry and humid conditions. It does not spread as rampantly as the normal *Hedera helix* (see page 158) which can often become a weed, so I prefer it for this reason. It is best trimmed flat once or twice a year to prevent leaders from mounding up. It also makes for a very handsome climber.

Hedera helix 'Pittsburgh'
Pittsburgh ivy

Mature height and spread: 4–10 m **Spacing**: 5 plants per m^2 **Plant form/habit**: Trailing **Position**: Full sun to full shade **Leaf hold**: Evergreen **Benefit**: Foliage **Climate**: Cold, cool temperate, warm temperate, subtropical, tropical **Water requirement**: Drought tolerant

This once widely used plant is now viewed as either a weed or a bit common, but I still like it for several reasons. You will find that the Pittsburgh variety with its finer, more divided leaf is not on the weed lists, but as these can change from area to area, check with your local authority before planting. It is without a doubt one of the most useful groundcovers for deep shaded areas and those with high root competition. It requires no watering and suppresses weeds incredibly well. It is not the most visually exciting of plants, I admit, so I save it for areas that are not highly visible or to stabilise banks or on retaining walls. I have also used it to great success when needing to hide ugly features such as bad rockeries. In this era of low supplementary watering, Pittsburgh ivy is often a great replacement for lawns and nature strips in particular, if local councils allow.

Ivy seems to grow in all areas, from tropical to cold. Just avoid planting ivy with exposure to the western sun, as the foliage may scorch in extreme heat and not repair until the following spring. Once established it needs no watering and very little maintenance. If it gets too high, simply brush-cut it to the desired height, but this may only have to be done once every three to four years.

Ivy puts on most of its growth as a spring flush, then continues to grow throughout the warmer months. To keep the plant within its boundaries, it can be pruned just after its spring flush for more formal gardens, or just before for a more relaxed check on the new growth. Throughout the warmer months, prune back growth leaders where they extend beyond their boundaries.

I also use Canary Island ivy (see page 184) extensively as a groundcover.

Opposite: *Hedera helix* 'Pittsburgh', *Quercus palustris, Buxus sempervirens, Hydrangea macrophylla, Syzygium australe*

Juniperus taxifolia 'Lutchuensis'
Creeping juniper

Mature height: 250 mm **Mature spread**: 3 m **Spacing**: 600 mm off-centre **Plant form/habit**: Dense spreading **Position**: Full sun, part shade **Leaf hold**: Evergreen **Benefit**: Foliage **Climate**: Cold, cool temperate **Water requirement**: Water until established

I am always looking for groundcovers that don't need pruning to keep them low and neat. This conifer is that plant. Its form is ground hugging and it will also spill over a wall and pot neatly without becoming too bulky. Its blue-green foliage is made of small needles and it sits beautifully with both green and silver foliage plants. It prefers full sun and requires free-draining soils. It will tolerate some humidity, making it perfect for use in Sydney and Brisbane as well as the cooler and drier states.

I use it en masse at the front of deep garden beds or as a deep border plant. It looks great under olives, as long as the olives are kept thin so as not to cause shading issues.

Rosmarinus 'Blue Lagoon' (syn. Salvie 'Blue Lagoon')
Blue lagoon rosemary

Mature height: 300–500 mm **Mature spread**: 1–2 m **Spacing**: 4 plants per m^2 **Plant form/habit**: Trailing **Position**: Full sun **Leaf hold**: Evergreen **Benefits**: Flowers, foliage, fragrance, culinary herb **Flowering time**: Jan–May **Flower colour**: Blue **Climate**: Cool temperate, warm temperate **Water requirement**: Drought tolerant

'Blue Lagoon' is the prostrate variety of rosemary I often use as a mass groundcover. Its flowers are small, but the foliage is aromatic when touched. It requires the same growing conditions as normal rosemary but instead of growing upright it spreads horizontally across the ground. It will also fall down a wall when planted on the top side of a retaining wall. It tends to flower more profusely than *Rosmarinus officinalis* (see page 132). It flowers heavily in winter and spot-flowers in January and February. I use it to cover large areas, especially in country gardens with expansive beds planted with nothing more than this plant and pencil pines (see page 280) for verticality and height.

This rosemary is well suited to coastal sites. It is more sensitive to humidity than its upright cousins. It likes well-drained soil and the only maintenance it requires is pruning to maintain height at approximately 500 millimetres – this can be done once every two to three years, although an annual clip encourages more young, vigorous growth. When pruning vast quantities, you can use powered hedge clippers to simply slash the foliage off.

Soleirolia solierolii (syn. Helxine solierolii)
Baby's tears

Mature height: Up to 300 mm **Mature spread**: 500–1000 mm **Spacing**: 9 plants per m^2 **Plant form/habit**: Trailing **Position**: Full shade **Leaf hold**: Evergreen **Benefit**: Foliage **Climate**: Warm temperate, subtropical, tropical **Water requirement**: Moisture-loving

Often you need a green carpet for areas where there is no sunlight, for example in light wells. Baby's tears is the perfect solution, and as long as it receives no direct sun it will thrive. It does, however, require a fair bit of moisture, and will not tolerate its root zone drying out. But if this happens it simply dies and comes back when the soil is sufficiently wet again. It remains very low to the ground, requiring no pruning to limit its height, and for this reason I often use it as an alternating pattern with cut stone or loose pebbles.

It will tolerate cold (but not frost), tropical conditions and heat as long as it is kept damp and in total shade. Its light green colour remains consistent throughout the year, making it a reliable plant for decorative purposes or simply as a carpet of green in a shady area.

The main drawback with this plant is its lifespan – I find it tends to die out in patches after several years. This may simply necessitate replacing the odd plant every now and then. If you are a keen gardener, you can propagate new material. Keep up a slow system of 'out with the old and in with the new'.

Thymus serpyllum var. albus
Creeping white thyme

Mature height: Up to 300 mm **Mature spread**: 300–500 mm **Spacing**: 9 plants per m^2 **Plant form/habit**: Trailing **Position**: Full sun to partial shade **Leaf hold**: Evergreen **Benefits**: Flowers, foliage, fragrance **Flowering time**: Nov–Jan **Flower colour**: White **Climate**: Cool temperate, warm temperate **Water requirement**: Drought tolerant

Thyme is a plant we have been using in our herb gardens as far back as gardening history stretches. This low-growing species is used in a decorative fashion. It is great between pavers to form a path or as a whole lawn area. It does like a sunny position, although I have grown it successfully in semi-shade. It absolutely needs to have well-drained soil, however – I often grow it in sand as this provides excellent drainage. It is drought tolerant and will thrive in extreme heat.

Thyme is an evergreen groundcover that remains very low, making it a great lawn replacement that requires no mowing. In the warmer months it is covered in tiny white flowers. It's happy in cold areas but not great with high humidity. It has an added bonus of a slight scent when trodden on or watered. It does, however, require some maintenance as weeds invade it easily. And as with baby's tears (see page 163), you might need to replace old plants every few years.

Trachelospermum asiaticum
Asiatic jasmine

Mature height: 0.4 m **Mature spread:** 1.5–2.0 m **Spacing:** 5 plants per 9 m^2 **Plant form/habit:** Groundcover **Position:** Full sun to part shade **Leaf hold:** Evergreen **Benefits:** Foliage **Flowering time:** Sep–Nov **Flower colour:** Cream to yellow **Climate:** Cool temperate, warm temperate, subtropical **Water requirement:** Moderate to dry tolerant

I have been using this groundcover form of the climbing *Trachelospermum asiaticum*, although that plant can be used effectively as a groundcover as well. The marked difference between the two species is the size of the leaf and the amount of flower produced.

T. asiaticum is better when you need a lower, more tightly hugging groundcover but will not reward you with prolific flowers or perfume. It's a very hardy evergreen groundcover and very effective when used en masse over large areas. It can be used as a low, dense mat of foliage, and looks at its best when clipped formally to retain uniform height. It is extremely hard to get it to flower but when it does, it will be covered in small, star-shaped, scented flowers in summer. Its main feature is its dark, glossy green foliage, with small-scale leaves suitable for small hedges or borders. I have successfully used it in very dry conditions as well as humid conditions.

Trachelospermum jasminoides
Star jasmine

Mature height: 500 mm (as groundcover) **Mature spread:** 2 m x 2 m **Spacing:** 750 mm or 5 plants per 2 m^2 **Plant form/habit:** Rambling groundcover **Position:** Full sun to partial shade **Leaf hold:** Evergreen **Benefits:** Flowers, foliage **Flowering time:** Dec–Jan **Flower colour:** White **Climate:** Cool temperate, warm temperate, subtropical **Water requirement:** Moderate

I've described the merits of this plant in the Climbers section (see page 200), however, I use it more extensively as a groundcover. Used en masse as a groundcover, it will form a wave of foliage to approximately 500 millimetres high and can be trimmed to maintain any height; if trimming, only prune after it has flowered, otherwise you will cut all the flower buds off. It is a very reliable and tough groundcover for dry or humid conditions. To train it as a groundcover, take the long, wispy stems it produces and pin them lightly to the ground for the first year. It will then look after itself, only requiring light horizontal trimming to the desired height. Select plants that are young and bushy, with supple leaders that will rest on the ground once growth thickens. It will grow in either shade or sun, making it very desirable for courtyard situations that often experience both conditions in a small space.

Vinca minor 'Alba'
White periwinkle

Mature height: 100 mm **Mature spread:** 1.5 m to infinite
Spacing: 1 plant per m^2 **Plant form/habit:** Groundcover
Position: Full sun to full shade **Leaf hold:** Evergreen
Benefits: Foliage and flowers **Flowering time:** Oct–Jan
Flower colour: White **Climate:** Cold, cool temperate, warm temperate, subtropical **Water requirement:** Drought tolerant

If you need a very low-hugging groundcover, this is the ideal choice. I use it when I need a tight mat of dark glossy green foliage under trees or tall shrubs, and it can also be used to spill over walls. It will take full sun or partial shade and will flower in both situations, although it flowers better in the sun. It does require some supplementary water in the summer months and will not tolerate wet conditions. It's a good weed-suppressing plant but tends not to affect trees that are planted amongst it. I use it under white crepe myrtle to great effect. Be wary as if kept unchecked it can take over the garden easily and is considered weedy in some areas.

Viola hederacea
Native violet

Mature height: Up to 300 mm **Mature spread**: 300–500 mm **Spacing**: 9 plants per m^2 **Plant form/habit**: Trailing **Position**: Full sun to full shade **Leaf hold**: Evergreen **Benefits**: Flowers, foliage **Flowering time**: Dec–April **Flower colour**: Blue **Climate**: Cold, cool temperate, warm temperate, subtropical **Water requirement**: Moisture-loving

It's great when you can use an Australian native plant, and this violet is one that I use extensively. It is a good choice for a filler for the front of garden beds or for between pavers in paths or terraces. It also works nicely to gap-fill the bare space that may exist at the foot of a hedge. A benefit is that it spreads rapidly and suppresses weed growth. It will tolerate some morning sun but will burn in extreme heat. It's actually very drought-hardy for a violet, and does not require much water. Small flowers cover the plant for up to six months of the year, but unlike its European cousins, it has little perfume to offer.

I have used native violet successfully in Melbourne, Sydney and even Brisbane. In the northern states I find it tolerates more sun.

It is particularly useful as a groundcover between flowers, creeping in between the spaces.

Climbers

Previous page: *Bougainvillea*
Above: *Wisteria sinensis*

When I was a child we lived next door to a very old estate that had the most magical overgrown orchard. Winding through some of the fruit trees was an old wisteria, and it was here that I fell in love with this beautiful climbing plant. Its perfume always signalled the start of spring and I was mesmerised by the way the flowers hung heavily from its branches. Ever since, wisteria, and climbers in general, have always reminded me of the happy childhood days I spent in that orchard.

No plant type is more critical to the successful blending of the house and garden than the climber. A newly built house can soon look more established and part of the landscape with a fast-growing climber trained to scamper up its walls. Everyone loves an instant garden nowadays, and that's what climbers offer.

One of the big benefits of climbers is that they are incredibly space efficient. They can rise to great heights without taking up too much ground space. This makes them ideal for courtyard situations where space is usually limited. Climbers can also assist with the temperature control of a house. Planting deciduous vines on a pergola adjacent to your home provides summer shade but allows winter sun to penetrate.

For me the biggest benefit that climbers provide is a softening factor. I like all my garden architecture to be draped in climbers. Perhaps Boston ivy (see page 193) on the walls, for example, and wisteria (see page 204) dripping over a portico. Obelisks covered in flowering roses (see page 196) and pavilions with lines of crimson glory vine (see page 203). These suit both formal gardens, where they can be used to soften straight lines in the design of the garden, and informal gardens, where they can be allowed to spill, drape and billow over the structure they're grown on.

The most important factor to consider when choosing a climber is whether it is to be evergreen or deciduous. Most people have a very definite preference

Above: *Trachelospermum jasminoides, Buxus sempervirens*

one way or the other. Some love deciduous climbers, enjoying the seasonal changes they offer, often with an abundance of flowers in spring or summer, good colour in autumn and bare vine in winter to let the sun shine through. Those who prefer evergreen climbers appreciate the lack of mess in autumn and the dependable green cloaking. Personally I prefer deciduous climbers because of the interest they provide through the year.

Climbers do require some work to train them to grow where you want, particularly when they are juvenile. They respond well if they are pruned from early on, just as a fruit tree does. Usually I allow only one main leader to climb vertically up whatever support they are to be grown on, and then allow branching to occur once it starts growing horizontally on the top of the structure. With wisteria I prune so that there is only one leader growing both vertically and horizontally, then allow the spurs to grow from the main leader.

Tips for climbers

1. Select the right climber for the right aspect. A bougainvillea (see page 177) will not be happy on the southern side of a structure, in full shade, whereas a creeping fig (see page 182) will tolerate it.
2. Time spent early on training and pruning a creeper will pay off quickly, usually within a year or two, with a well-shaped, healthy plant.
3. Some climbers, such as ivy and jasmine, can become invasive if left to their own devices. Cut back vigorously in spring before new growth occurs.
4. If selecting a deciduous climber, consider carefully the way it will look in each of the seasons.

Bougainvillea spp. and Cultivars
Bougainvillea

Mature height and spread: 3–5 m **Spacing**: 1 plant per post or 1 per lineal metre **Plant form/habit**: Climbing **Position**: Full sun **Leaf hold**: Evergreen **Benefits**: Flowers, foliage **Flowering time**: Nov–April **Flower colour**: Multiple **Climate**: Warm temperate, subtropical, tropical **Water requirement**: Moderate

No other climber creates more drama than bougainvillea, with its prolific growth habit and vibrant flowers. It provides a very strong statement in any garden. For me bougainvillea also creates a tropical or southern Mediterranean feeling. Thankfully this plant is one of few climbers that will tolerate a salt wind.

I have seen bougainvillea used extensively in Spain, southern Italy, the south of France, and even in California. In Australia we can pretty well grow it anywhere except in areas of extreme cold, where frost or snow occurs. I often use it for colour and I never shy away from the hottest shades of reds and magenta.

If you plant bougainvillea, use it confidently and exploit the drama it provides.

The downside of this climber is that it can often be slow or slightly temperamental to establish, especially in the southern states. It does require full sun to thrive and flower well; in fact, it will tolerate extreme heat with ease. It does have thorns, though, so it's not appropriate for households with small children. One other thing to remember is that possums seem to love it – so beware, a flourishing bougainvillea can quickly be destroyed if a possum is in residence nearby.

Bougainvillea is a robust climber and will require a very sturdy frame to grow on. I often use steel strapping for this purpose. Well-drained soil is a key factor in establishing young bougainvillea plants, and in cooler regions it is best to establish these plants over the warmer months.

Bougainvillea responds very well to pruning, and can be trained in the direction you desire. However, it can be hard to bring yourself to prune it over the summer because it means sacrificing the very reason that it is grown – the flowers!

Clematis spp. and Hybrid Cultivars
Clematis

Mature height and spread: 2–4 m **Spacing**: 1 plant per post or 1 per lineal 2 m **Plant form/habit**: Climbing **Position**: Full sun to partial shade **Leaf hold**: Deciduous **Benefit**: Flowers **Flowering time:** Sep–Feb **Flower colour**: Multiple **Climate**: Cold, cool temperate **Water requirement**: Moderate

Clematis is one of the more delicate climbers that are best grown by more experienced gardeners. It thrives only in the cooler states and will not tolerate any humidity.

If you live in a cold climate, prune most species, especially the large flowering hybrids, by cutting almost to the ground each year. If you live in a cool temperate region, simply pruning off old stems is sufficient to stimulate new growth. This all sounds a lot of work, but it will reward you with the most beautiful of climbing flowers. Clematis can also be planted in tandem with selected trees and shrubs to add flowers when the tree or shrub is not in flower.

There are two commonly grown types of clematis in Australia: the *Jackmanii* hybrid clematis and the montanas. The *Jackmanii* hybrids are the more responsive to pruning, and these should be cut down as specified by the producer. *Jackmanii* hybrids boast large, saucer-like flowers and come in a wide range of colours. I use them to mix with other climbers such as wisteria or roses, or I grow them on a less permanent stand in among herbaceous flowering plants. The spring-flowering montanas, on the other hand, are far more robust and don't require heavy winter pruning. For this reason they can be grown over a permanent structure such as a pergola or building. The beauty of this plant is that it can be almost left alone except for tidying up the dead wood after flowering.

Clematis will grow with partial shade but also does well in full sun. Some gardeners say it likes cool feet, so I keep the plant mulched and well fed. Clematis does require ample supplementary water. If grown on wires, high-tension stainless steel will suffice as clematis is not a very robust climber. But beware that possums seem to like eating it.

Above left: *Clematis* Jackmanii group (seed heads)
Left: *Clematis montana* type

Opposite: *Clematis* Jackmanii group (seed heads)

Clematis armandii
Clematis

Mature height: 9 m **Mature spread**: 6 m **Spacing**: 2 m **Plant form/habit**: Spreading climber **Position**: Full sun, part shade **Leaf hold**: Evergreen **Benefit**: Flowers **Flowering time**: Sep–Dec (spring to early summer) **Flower colour**: White **Climate**: Cold, cool temperate **Water requirement**: Moderate

This *Clematis* is one of the hardiest and easiest to grow of a genus that is often very temperamental. It is evergreen with long, glossy dark green leaves that hang vertically. The foliage is not its most beautiful feature, so I often mix it with wisteria or roses. It flowers prolifically in early spring with pinkish-white masses of star-like flowers. It requires a frame to grow over and will have to be trained for the first few years until it becomes attached to it. Try and keep the roots cool and moist and then the foliage will tolerate full sun or some light shade.

Ficus pumila
Creeping fig

Mature height and spread: 4–10 m **Spacing**: 2 plants per lineal metre **Plant form/habit**: Climbing **Position**: Full sun to full shade **Leaf hold**: Evergreen **Benefit**: Foliage **Climate**: Warm temperate, subtropical, tropical **Water requirement**: Drought tolerant

There is always great debate over the benefits of planting a creeping fig versus Boston ivy (see page 193). Both climbers have a self-clinging nature and the ability to cover large areas while remaining relatively flat against the structure they are climbing on. They differ in that only the climbing fig is evergreen, but a decision is usually easily reached when the gardener considers whether they want a deciduous or evergreen climber. However, Boston ivy grows much faster than the fig, and often speed is a critical factor. The fig is slow to establish and will often sit and appear to sulk for a year before really doing anything! However, once established, it grows moderately fast in full sun.

Creeping fig has two leaf forms: juvenile and mature, the former being the more desirable as it is a smaller leaf and remains flatter against its targeted structure. To retain the juvenile form, simply clip the plant back hard every second year.

Creeping fig is one of those rare climbers that will tolerate a range of conditions, from full shade to full sun, dry to humid. This makes it extremely useful for situations where you require consistency on all walls where some may face south or north. It does particularly well in dark areas and is one of the only plants to tolerate an enclosed light well or a dark, small courtyard. The only condition it will not tolerate is wet feet, so make sure adequate drainage is provided.

Creeping fig must be planted at the base of the structure you want it to climb. It will not grow in a downwards direction, so do not plant it at the top of a retaining wall, hoping it will creep over the wall below. Because the plant is self-clinging, it doesn't require any wires or supports. However, I have found that certain paints or lime washes prevent the plant from clinging for several years, so avoid planting near freshly painted walls. Creeping fig won't grow on Zincalume sheds either.

Hardenbergia violacea
Happy wanderer, false sarsparilla

Mature height and spread: 2–4 m **Spacing**: 1 plant per post or 1 per lineal 2 m **Plant form/habit**: Climbing **Position**: Full sun to partial shade **Leaf hold**: Evergreen **Benefits**: Flowers, foliage **Flowering time:** Sep–Jan **Flower colour:** Purple, white **Climate**: Cool temperate, warm temperate, subtropical **Water requirement**: Drought tolerant

One of our most attractive native climbers is the *Hardenbergia*. Its mass of deep purple flowers creates the impression from a distance of wisteria in full bloom. I also use it as a groundcover as it does equally well laterally. It tends to be a bit unruly as a climber, so don't expect it to behave and conform as a more formal climber would. It is extremely tough, though, and requires little to no supplementary watering.

This evergreen climber will grow in partial shade or full sun, and like most native plants, it doesn't like over-feeding. It's not self-clinging but will grow easily on tensioned stainless-steel wires. Don't mix it with other climbers, however, as it will take over.

'Happy Wanderer' is actually the registered name for a selected form of *Hardenbergia violacea* that has a particularly vigorous form, though the name is commonly used for all *Hardenbergia violaceas* in Australia. Another popular common name is false sarsparilla.

Aside from the usual purple variety, look out for two lovely white versions of this climber, 'Free and Easy' and 'Edna Walling Snow White'.

Above right: *Hardenbergia violacea*
Right: *Hardenbergia* 'Free and Easy'

Hedera canariensis
Canary Island ivy

Mature height and spread: 4–10 m **Spacing**: 1 plant per lineal metre **Plant form/habit**: Climbing **Position**: Full sun to full shade **Leaf hold**: Evergreen **Benefit**: Foliage **Climate**: Cold, cool temperate, warm temperate, subtropical **Water requirement**: Drought tolerant

Ivy has traditionally been one of our greatest climbers, probably due to its amazing toughness. It requires insignificant water if any, no feeding and very little maintenance except for an annual clip to keep it in check. It is, however, now considered a weed in many areas (although this often applies more to its cousin, the common ivy, *Hedera helix* – see page 158), so check first with your local authority.

I love Canary Island ivy for its large, glossy dark green leaf and its ability to hang like a curtain of greenery. It forms long runners that can hang vertically for up to 2 metres! It will self-cling given time, but does benefit from the use of a trellis structure. Like climbing fig it will tolerate full sun or full shade, so use it confidently in both situations.

Ivy creates a less formal look and so I use it when I want to design a more casual setting. To achieve this, allow it to remain relatively loose and hang like a curtain rather than keeping it tightly clipped.

Ivy will tolerate nearly all climatic conditions, thriving in both humid and cool climates. I have seen it growing in areas that receive heavy frost and remain unaffected. The only caveat is that it will burn if it is exposed to full sun above 40 degrees. So it is best to plant this species in easterly aspects or in part to full shade if you are in an area where temperatures exceed 40 degrees over summer.

Hydrangea anomala subsp. petiolaris
Climbing hydrangea

Mature height: To 15 m high **Spacing:** 1 plant per lineal metre **Plant form/habit:** Climbing **Position:** Part to full shade **Leaf hold:** Deciduous **Benefits:** Flowers, foliage **Flowering time:** November **Flower colour:** White **Climate:** Cold, cool temperate **Water requirement:** Moisture-loving

This climber is a little sensitive, but if you have the right conditions it will reward you with the most amazing show of delicate, lacy white flowers. Imagine having white, flat hydrangea flowers covering an entire wall of your house or garden wall; this is what will happen if you can grow this plant. It likes full shade or partial soft sun; try to avoid any hot summer sun. It likes moist soil and prefers a cool climate, although it will tolerate heat if in full shade and given ample water. Only try to grow it in mountain areas or the cooler southern states, and exercise patience as it grows slowly. It's self-clinging and will do little damage to your wall. Prune after flowering to remove stems which have flowered.

Jasminum azoricum
Summer jasmine

Mature height and spread: 2–4 m **Spacing**: 1 plant per post or 1 per lineal 2 m **Plant form/habit**: Climbing **Position**: Full sun to partial shade **Leaf hold**: Evergreen **Benefits**: Flowers, foliage, fragrance **Flowering time**: Oct–Mar **Flower colour**: White **Climate**: Cool temperate, warm temperate, subtropical **Water requirement**: Drought tolerant

Of all the jasmines, this is the least known, but it is no less beautiful for that. As the name suggests, it blooms in the heat of summer, filling the air with the sweetest of perfumes, often when all the other jasmines have finished flowering. The foliage is beautifully dark green and glossy, a perfect foil for the small but brilliant white flowers it produces. The flowers may be picked and brought inside to add fragrance to a room for a few days before they expire.

This climber provides an extended flowering season, unlike the Italian and spring jasmines that have main flourishes of flowers in spring and then nothing. Its growth habit is often sparse, making it acceptable to mix with other climbers, but not so suitable or attractive as a foliage plant on its own, in my view. This is probably why it is not as popular as other jasmines.

Summer jasmine is a relatively tough climber and will tolerate a few degrees of shade as well as full sun, but the more sun it receives the denser the foliage. I use it as a background climber that is meant to be smelt but not necessarily seen. It mixes successfully with roses, wisteria and clematis. It thrives in the southern states, as well as New South Wales and Western Australia. It tolerates some humidity, but not nearly as well as star jasmine (see page 200), so for this reason I would not use it in the tropics.

Lonicera periclymenum
European honeysuckle

Mature height: 3–4 m **Spacing:** 1 plant per post **Plant form/habit:** Climbing **Position:** Full sun to partial shade **Leaf hold:** Semi-deciduous (although may hold foliage in warmer climates) **Benefits:** Flowers (scented) **Flowering time:** Spring **Flower colour:** Pinkish red and cream to yellow **Climate:** Cold, cool temperate **Water requirement:** Moderate

Honeysuckle is a very old-fashioned climber that has largely disappeared from use in Australia. I have recently rediscovered it in an old country garden I was restoring. Its perfume is incredible and it will flower prolifically with absolutely no care. It does require wires on which to attach itself to a wall or pergola and will then billow out from the wall up to 500 millimetres. Mix it with roses or wisteria, or simply grow it on its own. I have mixed it through my hawthorn hedge with great success. It loves the heat and dry but will also tolerate some humidity. It thrives on neglect and requires no supplementary watering. Look for old-fashioned varieties, as these are the most perfumed. The most widely available variety is 'Firecracker' but it has little perfume. Prune back a third after flowering to encourage next season's flowers.

Parthenocissus quinquefolia
Virginia creeper

Mature height: To 15 m **Spacing:** 1 plant every 2 lineal metres **Plant form/habit:** Climbing **Position:** Full sun to full shade **Leaf hold:** Evergreen **Benefits:** Foliage **Climate:** Cold, cool temperate **Water requirement:** Moderate to drought tolerant

Parthenocissus quinquefolia has a large, dessiccated leaf and needs some support to get started on a wall but will eventually support itself as it is self-clinging. It is very vigorous and produces relaxed, strongly pendulous curtains of foliage. It has a more orange tint to its autumn foliage and will appear more sculptural on the wall. These are the two differences that set it apart from its cousin *P. tricuspidata*. I would use it if you want to soften a wall and prefer the orange tones in autumn. My strong preference is to use *P. tricuspidata* (see page 193), as it hugs the wall much more closely, and its glossier, fuller leaf is also more attractive. Best on cooler walls facing east and south but does tolerate full sun.

Parthenocissus tricuspidata
Boston ivy

Mature height and spread: 10+ m **Spacing**: 1 plant per post or 1 per lineal metre **Plant form/habit**: Climbing **Position**: Full sun to full shade **Leaf hold**: Deciduous **Benefit**: Foliage **Climate**: Cold, cool temperate, warm temperate **Water requirement**: Drought tolerant

This is one of my most-loved climbers as it so quickly converts a new garden into a more established one with its rapid growth habit. What other plant places its foliage in such an orderly fashion, leaf beside leaf, like a well-laid-out tiled wall? This ivy puts on a remarkable foliage display like no other. I am convinced that when Frank Lloyd Wright stated 'the job of a good landscape architect is to hide the work of a bad architect' he had Boston ivy in mind. This plant appears to engulf buildings and structures in front of your eyes, either hiding a multitude of sins or simply providing a sense of patina that usually only decades of establishment can provide.

Boston ivy is a self-clinging creeper that requires very little maintenance, except that you'll need to remove growth from parts of the building or structure where you don't want it to grow. I love this plant for its glossy, dark green leaf in summer and its brilliant, almost iridescent red in autumn. Even in the winter, its tracery of branches is attractive, particularly as the plant gets older.

I find Boston ivy requires no watering, and it will do well in either full sun or full shade. I use this species over its close relative *Parthenocissus quinquefolia* (see page 191), as the latter takes much longer to self-cling and does not form a dense cover of foliage so close to its structure. To keep Boston ivy regimented against a wall, make a small investment in time over the summer period to remove the lead growths that come out from the wall or structure that it climbs.

Opposite top: *Parthenocissus tricuspidata, Tilia, Trachelospermum asiaticum, Buxus sempervirens*

Rosa 'Albertine' and 'Wedding Day'
Climbing roses

Mature height and spread: 2–4 m **Spacing**: 1 plant per post or 1 per lineal 3 m **Plant form/habit**: Climbing **Position**: Full sun **Leaf hold**: Deciduous **Benefits**: Flowers, fragrance **Flowering time**: Nov–Dec **Flower colour**: Pale pink, white **Climate**: Cool temperate, warm temperate **Water requirement**: Drought tolerant

Traditionally climbing roses have been one of the main plants chosen as a climber, mainly for their incredibly beautiful flowers and perfume. They have for me long symbolised a style associated with pretty English gardens, and for this reason I have not selected them for my designs. They have, however, redeemed themselves throughout the very dry periods we have experienced in recent years. Roses love drought, and there are not many flowering plants we can say that about!

I've explored the use of a few varieties, and when choosing my favourites my requirements were for recurrent and robust flowering; the growth form being full and not too prolific; and not too delicate. 'Wedding Day' and 'Albertine' fulfilled these requirements for me in all ways except for the recurrent flowering.

'Wedding Day' is white with masses of clusters of flowers. Its exceptionally vigorous growth habit means it can cover pergolas or large structures within two years and then reward you with a mass of flowers in the spring. Be warned that it is very thorny, though. 'Albertine' is slightly more delicate in its growth habit and has very full pale pink cabbage flowers, plus a sweet perfume. Of course, there are many more climbing roses and I am starting to experiment with others.

As mentioned, roses require little watering and will not tolerate humidity, so they thrive in the southern states and love a hot, dry summer. They do need a lot of maintenance, however. The best roses are grown by maintaining vigour. Remove canes that are three years old or more, as this helps keep the wood young. Feed and prune regularly throughout the growing season to renew growth and remove tired stems. These are the best measures to fend off pests and diseases.

Left: *Rosa* 'Albertine'

Opposite: *Rosa* 'Wedding Day'

Rosa 'Pierre de Ronsard' PBR
Climbing Pierre de Ronsard rose

Mature height: To 3 m **Spacing:** 1 plant per 1 post **Plant form/habit:** Climbing **Position:** Full sun **Leaf hold:** Deciduous **Benefits:** Flowers **Flowering time:** Mostly in spring, but also summer and autumn **Flower colour:** Creamy white blushed with carmine pink to flower's centre **Climate:** Cold, cool temperate, warm temperate **Water requirement:** Drought tolerant; however, best to irrigate during extended dry periods

This rose will completely cover an entire wall and then reward you with masses of show-stopping mid-pink, full cabbage, lightly fragrant blooms in spring and summer.

It loves full sun and will tolerate dry conditions; it is no good in humidity but is otherwise vigorous and disease resistant. I have succumbed to its pinkness, using it on a north-facing wall in my rose garden at Stonefields. Even though it doesn't fit with the overall colour scheme, its abundance in growth and flowers makes it a must for any rose lover. Deadhead blooms to encourage repeat flowering.

This rose is a great picking rose, lasting for days as a cut flower.

Trachelospermum jasminoides
Star jasmine

Mature height and spread: 2–4 m **Spacing**: 1 plant per post of 2 per lineal metre **Spacing:** 2 plants per m^2 for a massed-planting effect **Position**: Full sun to full shade **Plant form/habit**: Climbing **Leaf hold**: Evergreen **Benefits**: Flowers, foliage, fragrance **Flowering time**: Oct–Feb **Flower colour**: White **Climate**: Warm temperate, subtropical, tropical **Water requirement**: Moderate

Star jasmine is one of the most versatile and hardiest climbers I use. It thrives in all conditions from heat to cold, from dry to humid, from full sun to full shade – whatever you throw at this plant, it seems to love it! As long as it receives some water in summer it can tolerate extreme heat – I am yet to see it burn. It's very versatile in that it can be used as a groundcover as well as a hedge climber to mass over a fence; not forgetting training it over pergolas, buildings and other structures. As a designer I find this plant invaluable and use it extensively.

In mid spring to summer star jasmine flowers profusely, providing masses of small star-shaped flowers with the most amazing cinnamon scent. This intoxicating perfume always reminds me of warm and humid Sydney summer nights. This plant's popularity started in New South Wales and has now spread throughout Australia.

This plant needs to be helped to climb initially, because its method of climbing is to twine around a structure. Help direct young leaders by tying them up to your support structure, then once the framework is covered, maintenance is just a matter of pruning back the plant to your desired shape. Star jasmine is also grown successfully as a topiary where it can fill a shape (a cone, for example) and be kept trimmed to this form.

I have trained star jasmine on wires to create various patterns, ranging from criss-cross diamonds to plain squares. You can keep it clipped relatively tightly against the wire, thus maintaining the pattern for years. Make sure your wires are strong enough and have a tension buckle on them, though, so you can retain the rigidity of the pattern. Better still, grow it on a fixed iron trellis to the desired shape.

I also like growing star jasmines side by side to create a hedge appearance. I find this method very useful for narrow side ways of houses where windows look out onto an unappealing paling fence.

Vitis coignetiae

Crimson glory vine, Korean grape

Mature height and spread: 4–10 m **Spacing**: 1 plant per post or 1 per lineal 3 m **Plant form/habit**: Climbing **Position**: Full sun to partial shade **Leaf hold**: Deciduous **Benefit**: Foliage **Climate**: Cold, cool temperate, warm temperate **Water requirement**: Moderate

Crimson glory vine is a type of ornamental grape, sometimes referred to as Korean grape. I am particularly attracted to this vine due to its large, textural leaf and its robust growth habit. I find it very architectural and bold, and for these reasons use it when I want to make a statement. Its leaves are deep green during the summer when the main attraction is its leaf size, and then in autumn they turn the most magnificent crimson, appearing from a distance to be on fire. When planted next to a maple that turns a similar shade of red, the combination is stunning and creates that sense of seasons that we often lack in Australia.

Being deciduous, crimson glory vine is great to grow on your house or an adjacent pergola as it provides wonderful summer shade, yet allows valuable winter light to penetrate. Like all true grapes, it only likes hot, dry summers and prefers a cool winter. Having said that, it does seem to cope with Sydney conditions. It tends to sit and sulk for the first year, but in subsequent years it grows with great vigour.

Keep it well pruned; unlike most other grapes, it can be pruned throughout summer without risk of exuding sap. Also, caterpillars do like this plant, so you'll need to keep them at bay.

Due to its vigorous growth habit this climber can be planted sparingly, with one plant able to reach at least 5–6 metres. But I still plant one plant per post if on a pergola as I love to see it twist its way up each post before spreading at the top.

A final word of warning: if you are going to incorporate this plant in your garden, be sure you source it locally. Grapes cannot be shipped around the country for fear of spreading the phylloxera aphid. Specimens should not be transported out of controlled areas.

Wisteria spp.
Wisteria

Mature height and spread: 4–10 m **Spacing:** 1 plant per post or 1 per lineal 4 m **Plant form/habit**: Climbing **Position**: Full sun **Leaf hold**: Deciduous **Benefits**: Flowers, foliage, fragrance **Flowering time**: Sep–Nov **Flower colour**: Lilac-blue to white **Climate**: Cold, cool temperate, warm temperate **Water requirement**: Moderate

Wisteria is probably my favourite climber – its long racemes can individually stretch to 1 metre in length, and together they hang like a curtain. For me wisteria is evocative of old English houses smothered in the plant after centuries of growth, and Italian gardens that have been allowed to decay in their own romantic way. It is a climber that responds perfectly to shaping, allowing the gardener to train it strictly over the lines of a house or even shaping it into a small tree. However, the strong vertical pull of each raceme calls for strategic placement around architecture. I most commonly use Chinese wisteria (*W. sinensis*) and Japanese white-flowered wisteria (*W. floribunda* 'Longissima Alba').

Wisteria is deciduous, which worries some people, but for me it could not be more appealing. Its bare branch structure in the winter holds up its spurs so proudly, it seems to be announcing that it will soon burst forth with the most beautiful of flowers. And the flowers are worth the wait. Not only are they stunningly beautiful to look at, they have the most intoxicating perfume. They also flower prolifically, with the entire climber smothered in flower.

Wisteria is easily grown and thrives on neglect – often in the worst of soils and with very little water! All it needs is lots of sun, so only grow it on the north or west side of a house, or if grown on a pergola or other structure, make sure it receives full sun. Wisteria will grow in partial shade but will not flower. I have seen excellent specimens in all capital cities, despite not thinking they would grow well in full tropical conditions. Place the plants roughly 2 metres apart for total coverage, and like the grape, plant one for each pergola post.

Train wisteria from an early age, allowing only one main leader to grow and then allowing flowering spurs to develop from each leader. The most admired wisterias are those that are tended to regularly with secateurs. Following the mass floral display in spring comes some sporadic flowering among the foliage and then the extension of the new growth leaders over summer. A regular visit to your wisteria with your secateurs is then necessary – prune these young new leaders back to 200–300 millimetres from the main leader, right through the summer. This will prevent it from becoming a tangled mess by the end of summer, and also help young wood to mature with flower-producing spurs in preparation for the spring display in the year to follow. Do not prune the spurs back too hard, though, as this will only stop the plant from flowering.

As with most climbers, wisteria in a pot will not grow to cover a good distance. If you have a balcony and you want it covered, plant the wisteria at ground level and allow it to grow up to the desired point.

Perennials and grasses

Previous page: *Verbena bonariensis*, *Nepeta* 'Dropmore', *Agastache mexicana* 'Sweet Lili'
Above: *Achillea millefolium* 'Terracotta'

The English idea of the perennial or herbaceous border was originally designed to provide the owner of a country house with a garden bed full of flowers that peaked in the summer months when the weather was warm enough for the garden to be in use. In Australia we are fortunate to have weather that allows us to enjoy being in our gardens for most of the year, and many of us are attracted to the idea of a perennial border, but of course we have a climate (or, I should say, climates, since our continent spans umpteen climate zones) that demands we use hardier plants that require less water.

From early on in the Australian colonies the trend became to include in a perennial border some flowering and non-flowering shrubs. More recently we have shifted to the style of using perennials with ornamental grasses that particularly suit the Australian climate. And modern-day perennial hybrids have also extended the flowering life of the border; this means it is possible to have a garden in flower for over six months of the year.

Perennials have also proved to be extremely drought hardy. One of the most famous 'dry' perennial gardens can be found at Lambley Nursery and Gardens in Ascot, Victoria. The perennials in this stunning garden are deep-watered only three times a year. This and other similar gardens have set the tone for plant selection styles that are cropping up all over the southern states of Australia.

I particularly love using perennial borders in country gardens or very large city gardens. My own new rural garden at Stonefields has many herbaceous borders and I use these to experiment constantly with new perennial plants. For perennial borders in large gardens, I use perennials and grasses only. For smaller city gardens it is a good idea to integrate small shrubs into the border. This allows for the perennials to die down in the winter and any holes to be disguised by overhanging or nearby shrubs.

Perennials and grasses provide the garden creator with a type of plant that will provide flowers in the garden for a lengthy period of time, more so than any other. They are a key plant group that brings an ephemeral quality to the garden and that I always think of as being light in mass and therefore able to weave easily among the fabric of other, heavier plant types.

Creating the perfect herbaceous border is a true art and one that I have learnt requires much practice. In my country garden at Stonefields I have reworked my

Above: *Iris sibirica*

perennial borders every year since they were first created sixteen years ago. And I'm not done yet. The height that a plant grows to can have an unexpected effect on the look of the whole, and I am always removing certain clumps of perennials or grasses and placing them in another position in the border. I am also not afraid to throw plants out altogether. You can't get sentimental about plants if they don't work.

I used to believe in layering plants with the shortest at the front of the border and the taller at the rear, but I have changed my opinion on this now. I like a herbaceous border to be tall right to the front, with the first layer only marginally shorter than the rear. I place my yarrows (see page 212) and sedums (see page 258) at the front and use taller grasses such as *Calamagrostis* × *acutiflora* 'Karl Foerster' (see page 222) at the rear. I like the border to be visually airy and light and not weighed down with perennials that are too bulky. For this purpose I love the plants that have minimal foliage and hold their flowers high on long stems, such as yarrow and *Allium* 'Purple Rain' (see page 215).

Tips for perennials and grasses

1. When planting out ornamental grasses, don't overcrowd them. Allow them to breathe, so that they can move in the wind.
2. Perennials need a deep bed to be successful. Narrow beds do not allow for enough mass and interest. My minimum depth for a good perennial bed is 3 metres. It is ideal to have 5 metres to work with.
3. Consider layers when designing a perennial bed. Research the mature heights of all selected plants and do not be afraid to place plants of medium height at the front of the bed.
4. Mound the soil in preparation for planting out a bed so that it is 400–500 millimetres higher than the surrounding lawn or paving. This helps with drainage.

Achillea millefolium 'Terracotta'
Yarrow

Mature height and spread: 300–500 mm **Spacing**: 4 plants per m^2 **Plant form/habit**: Clump-forming **Position**: Full sun **Leaf hold**: Evergreen **Benefits**: Flowers, foliage **Flowering time**: Dec–April **Flower colour**: Yellow to terracotta **Climate**: Cold, cool temperate, warm temperate **Water requirement**: Drought tolerant

Achilleas are one of my favourite herbaceous perennials and 'Terracotta' is a stand-out variety for me. It starts out yellow in colour and then fades to a wonderful dry shade of terracotta. This plant draws attention to itself by the shape of the flower heads. Many tiny flowers are arranged into a flat umbrella and together these coloured panicles are held approximately 500 millimetres above the foliage in an eye-catching display. I often use other varieties too, mainly to introduce yellow into a border. They flower in mid- to late summer, and if cut back you can often get a second flush of flowers in autumn.

The plant is not self-supporting and tends to collapse under the weight of all its flowers. To avoid this I weave twigs from old hawthorn prunings into a basket-like structure to support the tender stems and simply push it into the ground in winter, over the plant. In England they use prunings from hazelnut trees for this purpose, so I have planted several hazelnuts with a view to trying this. Lift and divide the clump every few years as this helps in flowering but also prevents it taking over the entire bed. Discard approximately half the plant and replant the rest.

Full sun and well-drained soil is necessary for yarrow. The plant will not tolerate humidity. I position yarrow towards the front of the garden bed, if not at the very front.

Be aware that in some areas *Achillea* is considered a weed, so check carefully with your local authority before planting.

Agastache 'Blue Fortune'

Giant hyssop

Mature height: 1 m **Mature spread:** 600 mm **Spacing:** 5 plants per 3 m^2 **Plant form/habit:** Upright **Position:** Full sun **Leaf hold:** Deciduous **Benefits:** Flowers, aromatic foliage **Flowering time:** Summer to autumn **Flower colour:** Violet–blue **Climate:** Cold, cool temperate **Water requirement:** Moderate

This is one of the best blue-flowering *Agastache*; given the right conditions it will successfully flower for 4–5 months of the year throughout late spring–summer and into autumn. It's a very welcome addition to any perennial border, doing well in the middle section of the border. The spikes of vibrant blue flowers provide accents, standing out like beacons, and are attractive to bees. It needs well-cultivated and drained soil rich in compost. It doesn't do well in sandy soils or humid conditions. You need do nothing to this plant except cut it to the ground in winter and compost over.

Agastache 'Sweet Lili'
Mexican hyssop

Mature height and spread: 500–1000 mm **Spacing**: 1 plant per m^2 **Plant form/habit**: Upright **Position**: Full sun **Leaf hold**: Deciduous **Benefits**: Flowers, foliage, fragrance **Flowering time**: Oct–May **Flower colour**: Purple-blue **Climate**: Cool temperate, warm temperate **Water requirement**: Drought tolerant

This bushy, fast-growing hyssop produces long-lasting flowers on spikes from early summer to winter. Its foliage is aromatic when crushed. I use it in all my perennial borders as it is such a long-flowering plant and has a very distinct and unusual flower colour. It is almost a soft lilac blue and very dusty in appearance, which becomes the perfect foil for all the other shades, including the more vibrant hues of blue, in the border. It flowers on the very end of its stems, creating a wonderful wave of flowers when there is a breeze.

This hyssop grows to approximately 1–1.2 metres in height, so plant it towards the rear of the bed. It loves a dry hot summer and requires very little water. Like most herbaceous perennials, it will not tolerate any humidity. Hyssop should be cut completely to the ground every autumn, so it will then re-shoot in the spring. It tends to be the last plant in the border to finish flowering. It may need renewal every four to five years by taking new slips from the edges of the plant.

Allium 'Purple Rain'
Ornamental onion

Mature height: 900 mm **Mature spread**: 200 mm **Spacing**: 500 mm **Plant form/habit**: Bulb/rosette **Position**: Full sun **Leaf hold**: Evergreen **Benefit**: Flowers **Flowering time**: Dec **Flower colour**: Purple **Climate**: Cold, cool temperate **Water requirement**: Low

Allium 'Purple Rain' is a form of *Allium* I have found much easier to grow and maintain from year to year than its more temperamental relative *A.giganteum*. It is shorter in height and stature but no less exciting with its brilliant purple globe-shaped flowers. If you are one of the few lucky people who can grow both *Alliums*, plant them together for an extended flowering period. *Allium* 'Purple Rain' will flower in mid spring and will then be followed by *A. giganteum*.

The bulbs may need to be lifted every two to three years and divided. Plant them twice as deep as the size of the bulb and they should easily survive the summer in the ground if you don't overwater the garden bed. I grow them in my flowering herbaceous perennial beds – they emerge and flower before the perennials amass too much foliage following their winter cut down.

Anemone × hybrida
Japanese windflower

Mature height: 500–1000 mm **Mature spread**: 300–500 mm **Spacing**: 4 plants per m^2 **Plant form/habit**: Clump-forming **Position**: Partial shade to full shade **Leaf hold**: Evergreen **Benefits**: Flowers, foliage **Flowering time**: Feb–Apr **Flower colour**: White, pink **Climate**: Cold, cool temperate, warm temperate **Water requirement**: Moderate

For a flower that easily spreads throughout the garden without becoming a pest and taking over, this plant cannot be overlooked. Its very delicate foliage and equally delicate flowers appear to float on long stems like butterflies in the garden. For most of the year the foliage is low and inconspicuous, which allows it to weave in and out of other plants without dominating the bed. You can even forget that you have Japanese windflower in your garden until early autumn when every plant sends up a flower stem in a rewarding display that reminds you of this plant's valuable contribution to the semi-shaded garden.

Japanese windflower boasts flowers of varying shades of pink and white, but I prefer the white as I find the pink a bit 'lolly' in colour.

This plant has been burning badly in the heat of recent summers so grow it in the shade or in morning sun only. It likes cool and wet conditions, so make sure you have an adequate water supply. It grows well in Sydney but I would not subject it to much more humidity than this. In cooler regions, cut it to the ground in winter. It does well in frost and snow and is pest- and disease-free.

Anthemis tinctoria × punctata subsp. cupaniana 'Susanna Mitchell'
Flowering chamomile

Mature height: Up to 300–500 mm **Mature spread**: 1–2 m **Spacing**: 1 plant per 1.5 m^2 **Plant form/habit**: Spreading **Position**: Full sun to partial shade **Leaf hold**: Deciduous **Benefit**: Flowers **Flowering time**: Dec–May **Flower colour**: Cream **Climate**: Cold, cool temperate, warm temperate **Water requirement**: Moderate

Anthemis is a groundcover herbaceous perennial and is useful for the very front of a flowering border or used en masse to line a path. Being herbaceous it does need cutting back in the winter if you live in a cold or cool climate, so remember this when choosing its position. It is extremely fast growing and can tend to take over, so be aware. It has a pale cream flower that is abundant from early summer right through to late autumn. Few plants flower for so long. It does not need staking but will require moderate watering. It likes full sun but will tolerate some shade, where it will just not flower as well.

It is best to lift and divide this plant every year or two to keep the plant tidy, to maintain vigour, to keep it within its boundaries and to help prolong its season in bloom. Each plant will spread approximately 1.5 metres, so position to allow for this spread, which will be achieved in a matter of months.

Arthropodium cirrhatum
New Zealand rock lily

Mature height and spread: 500–1000 mm **Spacing**: 4 plants per m^2 **Plant form/habit**: Clump-forming **Position**: Full sun to full shade **Leaf hold**: Evergreen **Benefits**: Flowers, foliage **Flowering time**: Jan–Mar **Flower colour**: White **Climate**: Cool temperate, warm temperate, subtropical **Water requirement**: Drought tolerant

New Zealand rock lily is a plant I have only discovered in the past few years, mainly through researching plants that do well in drought conditions. It has been an invaluable discovery, not only because it is drought-hardy but also as it does well in heavily shaded areas where root competition is high from major trees.

It has fleshy, strappy leaves and attractive sprays of white flowers that appear in late spring and summer. I like to use it in large drifts in the garden and I find it works well alongside drifts of clivia (see page 358).

It really needs very little maintenance, but if the leaves become scrappy, cut the plant to the ground and it will re-shoot all new foliage. If this plant has access to some soil moisture and compost, the floral display above the foliage is vastly improved.

It will also take some sun but does not like afternoon or extreme summer sun. It will tolerate some humidity, though I do find the more humid the environment the more prone it is to pests, in particular caterpillars.

Be vigilant for snail and slug activity, too, because these pests can ruin the pure elegance of this plant for months.

Aster × frikartii 'Mönch'
Frikart's aster

Mature height and spread: Up to 500–1000 mm **Spacing**: 4 plants per m^2 **Plant form/habit**: Rounded **Position**: Full sun **Leaf hold**: Deciduous **Benefit**: Flowers **Flowering time**: Feb–May **Flower colour**: Blue **Climate**: Cold, cool temperate, warm temperate **Water requirement**: Moderate

Frikart's aster is the best of the aster daisies. Forming a neat sphere shape, this plant remains low and becomes smothered in blue flowers in late summer and into autumn. It works well when placed in the front of the border where its flowers are much appreciated in late autumn but also its shape and foliage are still interesting while not flowering. The intense blue flower attracts much attention, so place it where you want the eye to be led. I often use this plant in beds with other purple-flowering perennials.

Towards the end of its flowering, cut it back and it will provide you with another flush of flowers. It is self-supporting and herbaceous, so cut it to the ground in early winter.

It likes a cold winter and dry heat in summer, but will not tolerate humidity.

Opposite: *Aster × frikartii* 'Mönch', *Agastache* 'Blue Fortune', *Agastache* 'Sweet Lili'

Calamagrostis × acutiflora 'Karl Foerster'
Feather reed grass

Mature height and spread: 1–2 m **Spacing**: 1 plant per m^2 **Plant form/habit**: Grass **Position**: Full sun **Leaf hold**: Evergreen **Benefits**: Flowers, foliage **Flowering time**: Mar–Sep **Flower colour**: Mauve turning yellow **Climate**: Cold, cool temperate, warm temperate **Water requirement**: Moderate

I know ornamental grasses are fashionable right now, but for good reason: they are tough and I believe they bring a border to life. *Calamagrostis* is a stand-out plant that makes a huge statement when planted en masse. The mauve feathery flower heads are held well above the foliage and they age to a straw yellow in the autumn and hold right through to spring, providing a stark contrast to the evergreen mass of foliage below. I have *Calamagrostis* in a border in my own garden and it is the first to flower and the last to finish. In fact, it doesn't finish – we leave the dry flower heads on as they look great when covered in frost. It looks great in a breeze when the wind allows the flower heads to dance in the border. I cut them down just before spring when they are about to shoot again.

'Karl Foerster' grows to about 1.2 metres high, so I recommend placing it at the back of the border. It also combines well with other grasses if you want to create a bed of grasses only. It is self-supporting, so does not need staking.

Like most perennials it thrives with a cold winter and hot dry summer. It does not need excessive watering but does appreciate some summer moisture. Keep the plants fresh and vigorous by dividing grass clumps once every two years.

Below: *Calamagrostis × acutifolia* 'Karl Foerster', *Achillea millefolium* 'Terracotta', *Sedum* 'Autumn Joy'

Erigeron karvinskianus
Seaside daisy

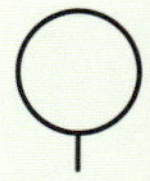

Mature height and spread: 300–500 mm **Spacing**: 6–8 plants per m^2 **Plant form/habit**: Rounded **Position**: Full sun to partial shade **Leaf hold**: Evergreen **Benefits**: Flowers, foliage **Flowering time**: Oct–May **Flower colour**: White turning pink **Climate**: Cool temperate, warm temperate **Water requirement**: Drought tolerant

The seaside daisy is the perfect plant for softening hard surfaces. Edna Walling used it extensively, growing it out of cracks in the paving; I plant it at the base of walls and find it especially handy for the footpath side of front boundary walls where space is often limited. This plant thrives in tight spaces, growing in beds as narrow as a couple of centimetres in width. It seems to grow anywhere, in shade or full sun, although it becomes leggy in full shade. It is often associated with cottage-style gardens due to its softness and informal growth habit. Its small daisy-like flowers start out white and age to a dusky pink.

The seaside daisy requires little water and will tolerate poor soil conditions though it does not like having wet feet. In fact, in some areas it grows so successfully it is considered a weed, so check with your council before planting. To prevent it from becoming woody or too leggy I cut it to the ground at least three to four times a year, letting it flower for a few months and then cutting it back hard. This keeps the plant looking fresh and flowering for the entire year. Plant from tube stock – it grows so fast there is no need for larger pot sizes.

Euphorbia amygdaloides var. robbiae
Wood spurge

Mature height: 500 mm **Mature spread:** 500+ mm **Spacing:** 4 plant per 1 m^2 **Plant form/habit:** Moderately spreading (by rhizomes) **Position:** Part shade to full shade **Leaf hold:** Evergreen **Benefits:** Foliage, flowers **Flowering time:** Spring **Flower colour:** Chartreuse, becoming autumnal as the weather cools **Climate:** Cold, cool temperate, warm temperate **Water requirement:** Drought tolerant

This is a great low-growing, spreading *Euphorbia:* dark green with textural interest, and a tight rosette of foliage carried by every upright stem. Unlike other *Euphorbia* spp., it is not likely to simply die for no apparent reason. I use it almost as a tall groundcover, allowing it to form a large mass planting in semi-shaded situations. It seems to do well under trees where it's hard to get any plant established. It will take full sun but it does best for me in a woodland situation. The lime-green flowers are vibrant and add an electric feel to the garden when in bloom, illuminating the area they are planted in. Cut them down to a low mat in the winter.

Euphorbia characias subsp. wulfenii
Euphorbia

Mature height and spread: 1–2 m **Spacing**: 2 plants per m^2 **Plant form/habit**: Rounded **Position**: Full sun to partial shade **Leaf hold**: Evergreen **Benefits**: Flowers, foliage **Flowering time**: Aug–Jan **Flower colour**: Lime-green **Climate**: Cold, cool temperate, warm temperate **Water requirement**: Drought tolerant

I love the colour green in the garden and in particular green flowers. Sedums (see pages 258–59) provide me with this display, and the other genus is *Euphorbia*. This is a huge group of plants and I tend to use a diverse range of them, but my favourite is *Euphorbia characias* subsp. *wulfenii*. It flowers for a long period, up to six months of the year, and it provides height and form with its clusters of green, bell-shaped flowers. It can be temperamental, though; appearing to thrive and then one day dying for no apparent reason. I nearly always put this down to drainage. It loves dry, open well-drained soil and full sun. If your soil is too heavy, add coarse sand and mound the soil so it drains well, and then euphorbias should thrive.

The acid-green flowers are a good combination with *Echium candicans* (see page 362), with which they share the same flowering period. Once the flowers have faded and new growth has started to emerge from within the plant, cut out every tired flower stem right to its base and allow the new foliage to recreate a new and tidy plant for the year to follow. Wear gloves and eye protection because the milky sap that is released during pruning is very irritating when you come into contact with it.

Euphorbias do not tolerate humidity well, so plant them only in the drier southern states. They do have a very limited life, so expect no more than three to four years from them and you will not be disappointed. They are great companions to the sedums, so plant them in combination for a rewarding show.

Gaura lindheimeri 'Whirling Butterflies'
White gaura

Mature height: 1000 mm **Mature spread:** 750 mm **Spacing:** 5 plants per 2–3 m^2 **Plant form/habit:** Upright to irregular **Position:** Full sun to partial shade **Leaf hold:** Semi-deciduous **Benefits:** Flowers **Flowering time:** Nov–Apr **Flower colour:** White with red sepals **Climate:** Cold, cool temperate, warm temperate **Water requirement:** Drought tolerant

This plant is one of the best perennial companions to roses and other shrub gardens. The foliage is relatively insignificant but its long, waving spires of flowers will rise up between the flowering shrubs and add another layer of flowering. It is almost indestructible and will flower for well over five months. It is a pure white cultivar with green stems. I also use it in my perennial borders, where it squeezes in easily between denser foliage perennials, revealing no more than its tall flowering spires of white flowers. It loves cold winters and dry summers but will tolerate some humidity. Cut down to approximately 200 mm of woody stem in winter. One of the most reliable flowering and growing perennials.

Geranium Rozanne® 'Gerwat'
Cranesbill

Mature height: 500 mm **Mature spread:** 600 mm **Spacing:** 5 plants per 1–2 m^2 **Plant form/habit:** Clump-forming **Position:** Full sun to partial shade **Leaf hold:** Deciduous **Benefits:** Flowers **Flowering time:** Oct–Apr **Flower colour:** Light blue with white centre **Climate:** Cold, cool temperate, warm temperate **Water requirement:** Moderate

The species geraniums are one of the most delicate and beautiful of the perennials, and Rozanne is the best of them. They have a reputation for being slightly delicate but I find Rozanne to be very hardy and easy to grow. Its mid-blue flowers come in spring and will continue well into summer. They are the perfect height to grow under roses and are a perfect companion to any rose garden. They also sit well in any perennial border or used en masse as block planting. I grow them in full sun but they do benefit from a little shade or protection from the elements provided by larger plants. After their initial spring flush of flowers, remove all dead flowers and minimal foliage and they will return back into flower in a matter of a few weeks.

Helleborus argutifolius
Corsican hellebore

Mature height and spread: 300–500 mm **Spacing**: 8 plants per m^2 **Plant form/habit**: Clump-forming **Position**: Partial shade **Leaf hold**: Evergreen **Benefits**: Flowers, foliage **Flowering time**: Aug–Oct **Flower colour**: Lime-green **Climate**: Cold, cool temperate, warm temperate **Water requirement**: Moderate

Hellebores are old-fashioned plants but have remained popular, probably due to their hardiness. They prefer shade but will tolerate some sun, preferably not afternoon sun. They flower in winter and early spring when the garden is often light on for flowers.

They are typically a woodland plant and therefore thrive under most deciduous trees. They grow well in the southern states and even in Sydney, although not much further north than this. They do like moist conditions but will tolerate some degree of dryness during summer. Throughout the drought we recently experienced most of the hellebores survived without supplementary water. They excel in cold climates and do well in both frost and snow.

I use them almost as a mass groundcover under trees in a woodland situation. They also look equally interesting as specimen plants dotted among other shade-loving plants.

This species is the toughest of the hellebores, tolerating more sun than the other species and requiring less water. If you are desperate for a *Helleborus* and only have sun and not much water, this is the plant for you.

This hellebore can be grown for its leaf texture alone. Each leaf has three thick, leathery leaflets with toothy leaf margins. The leaves are visually strong and coarse to touch. The leaves are a quiet dull pale green, while the flowers are lime-green cups held above the foliage in a subtle display.

Helleborus foetidus
Stinking hellebore

Mature height and spread: 300–500 mm **Spacing**: 5 plants per m^2 **Plant form/habit**: Clump-forming **Position**: Full sun to partial shade **Leaf hold**: Evergreen **Benefits**: Flowers, foliage **Flowering time**: Aug–Oct **Flower colour**: Lime-green **Climate**: Cold, cool temperate, warm temperate **Water requirement**: Moderate

This particular hellebore I use for the unusual sculptural shape of its leaves. Instead of being rounded like the *Helleborus orientalis*, they are more dissected, forming a fan shape. This makes for a more interesting plant while not in flower.

This hellebore flowers with groups of small lime-green cups presented on strong, thick stems held well above the dark green foliage of the main plant below. A short-lived plant, it will, however, re-seed.

Cut away old stems when new growth pushes through to keep the plant looking young and tidy.

Helleborus orientalis
Lenten rose

Mature height and spread: 300–500 mm **Spacing**: 5 plants per m^2 **Plant form/habit**: Clump-forming **Position**: Partial shade to full shade **Leaf hold**: Evergreen **Benefits**: Flowers, foliage **Flowering time**: Jun–Sep **Flower colour**: Multiple **Climate**: Cold, cool temperate, warm temperate **Water requirement**: Moderate

This is my very favourite hellebore. It is also the most delicate. The Lenten rose flowers more prolifically and spreads more quickly than other hellebores. They come in a range of colours from black to pink to whites and creams. Plant them in deep, composted soil. Mulch well, using well-rotted oak leaves if you can find them.

To keep plants tidy, it is good to remove the older leaves around late autumn. The flowers are mostly nodding and so the plants ask that you bend down and have a look into the flower to reveal their real beauty. Young, fresh blooms can be appreciated further by bringing some inside and floating them in a bowl of water on a table.

I use Lenten roses en masse, particularly in semi-shaded wooded areas. It will tolerate some sun but prefers shade.

Hemerocallis cultivars
Daylily

Mature height: 500–1000 mm **Mature spread**: 300–500 mm **Spacing**: 6 plants per m^2 **Plant form/habit**: Strap leaf **Position**: Full sun to partial shade **Leaf hold**: Evergreen (preferred) **Benefits**: Flowers, foliage **Flowering time**: Oct–May **Flower colour**: Multiple **Climate**: Warm temperate, subtropical **Water requirement**: Drought tolerant

Daylilies bring a flower border to life with their brightly coloured flowers. The flowers rise above the leaves, presenting themselves on the ends of long stems in a series of large star-shaped blooms. I love them for their flowers rather than their foliage, which is a rather dull collection of strap-shaped leaves.

I find daylilies slightly temperamental to grow. They can flower well for the first season and then not as well in following seasons. The secret seems to be to feed them regularly with compost and liquid food and to cut them back after flowering.

Daylilies tolerate humidity and seem to thrive in Sydney and do moderately well in Brisbane. In the southern states, be careful of frost, as this can burn the foliage.

There are so many varieties of daylilies that it can be difficult to choose one colour alone. Select a cultivar that is a suitable size, be it dwarf, medium or tall. And stick to the evergreen varieties, as those that are completely dormant will leave bare patches in the garden over the winter.

Iris germanica
Bearded iris

Mature height: 500–1000 mm **Mature spread**: 300–500 mm **Spacing**: 6 plants per m^2 **Plant form/habit**: Strap leaf **Position**: Full sun **Leaf hold**: Evergreen **Benefits**: Flowers, foliage **Flowering time**: Oct–Dec **Flower colour**: Multiple **Climate**: Cool temperate, warm temperate **Water requirement**: Drought tolerant

I used to believe that irises were only good for their amazing display of flowers in late winter/early spring. Certainly this is a very good reason to plant them, but I have since come to appreciate their foliage either as a grey–green addition to a mixed planting scheme or as a mass planting in a narrow bed adjacent to a wall or path.

Irises are versatile plants that fit into many different planting schemes. They are equally at home in a flowery English-style border, in a dry Mediterranean-style garden, or en masse in a contemporary design. They often flower when little else is out in the garden and come in a huge range of colours. The colour range is one of the reasons I use iris. Often I am able to match interior fabric colours to an iris colour planted in view of a sitting room.

Irises like dry, warm and extremely well-drained soil. If your soil is not well-drained, add coarse sand and raise the bed to prevent water sitting around their roots.

Until recently irises were always sold as bare-rooted rhizomes, but now they seem to be sold in pots as well. Either way, place them in a circle of three or five to form a clump among other plants. If you are planting iris as a single species per bed, plant them fairly close together, approximately 200 millimetres apart.

Cut down the flower stems once flowers have finished in early summer and tidy up the plants during the winter by removing old and tired foliage. Be aware also that they may need lifting and dividing every four to five years. The maintenance regime is utterly worth it for this very rewarding plant.

Iris germanica is the hardiest of the irises and a good bet if you don't have the water supply needed to keep other species healthy and happy.

Opposite: *Iris germanica, Lavandula dentata,* David Austin rose

Iris sibirica
Siberian iris

Mature height and spread: 500–1000 mm **Spacing**: 6 plants per m^2 **Plant form/habit**: Strap leaf **Position**: Full sun **Leaf hold**: Deciduous **Benefits**: Flowers, foliage **Flowering time**: Nov–Dec **Flower colour**: Multiple **Climate**: Cold, cool temperate **Water requirement**: Moderate

If *Iris germanica* (see page 234) is the toughest of the irises, then *Iris sibirica* is one of the more sensitive. It needs far more water and richer soil, and appreciates the cool weather, although it does tolerate the heat. It may even tolerate wet soil far more than *Iris germanica*.

This iris flowers only once with one flush in spring, but it is well worth it as the show is spectacular. This is a tall iris, reaching a metre and more, especially when in flower, so place it either one per corner of the bed or towards the centre.

The flowers come in many colours, but I prefer the shades of blues and whites. Give them plenty of water and compost and cut them to the ground every winter and you will be rewarded with the most amazing display of flowers, albeit only for three weeks or so. Believe me, it is well worth it. And if you want to mitigate the disappointment when your *Iris sibirica* plants stop flowering, simply grow a range of irises so that as one group finishes another begins.

Note that with this iris, one plant tends to form a decent clump, so they don't need to be so densely planted.

Lilium auratum and L. speciosum Hybrids
Oriental lily

Mature height: 1–2 m **Mature spread**: 300–500 mm **Spacing**: 5 plants per m^2 **Plant form/habit**: Upright **Position**: Full sun to partial shade **Leaf hold**: Deciduous **Benefits**: Flowers, fragrance **Flowering time**: Jan–Feb **Flower colour**: Multiple **Climate**: Cold, cool temperate, warm temperate **Water requirement**: Moderate

I often use oriental lilies as fillers in a flower bed. Their large blooms are available in varying shades of pink, yellow and white. And the flowers possess beautiful shapes, sometimes resembling deep funnels or large stars. Deliciously fragrant, these flowers will never disappoint.

These lilies are so narrow that they can fit among other clumps of plants and simply provide flowers when they are required. I tend to cluster them in groups of three, planting them approximately 300 millimetres apart. I place them randomly in the border, regardless of other plants, allowing them to appear anywhere from the centre to the rear of the bed.

Although these lilies like full sun, they will tolerate partial shade. Rich, well-drained soil is important, so add plenty of compost and coarse sand. Buy them as bulbs for planting in the autumn.

I seem to be able to leave the bulbs in the ground and they still reappear every year. If your soil is not well drained, it may pay to lift them after they die down and replant in the autumn.

Note that they are prone to collapsing, so I recommend staking them with bamboo canes as they grow.

Limonium perezii
Sea lavender

Mature height and spread: 300–500 mm **Spacing**: 4 plants per m^2 **Plant form/habit**: Clump-forming **Position**: Full sun **Leaf hold**: Evergreen **Benefits**: Flowers, foliage **Flowering time**: Dec–Mar **Flower colour**: Blue **Climate**: Cool temperate, warm temperate **Water requirement**: Drought tolerant

During the drought, I discovered that this plant is one of the toughest perennials – it really thrives on neglect. It is only short, with its foliage typically reaching just 200–300 millimetres, and its flowers are carried above this in summer with a spray of deep lavender-blue flowers.

Sea lavender mixes well with succulents, providing the flowers that a drought-tolerant garden often lacks. The plant forms a clump very rapidly, so position sparingly and allow for it to spread.

I have grown sea lavender in all the southern states as well as in Sydney. As its name suggests, it is also a great coastal plant, growing well in sandy soils – its leathery leaves tolerate salt winds. It likes full sun and well-drained soil.

The blue colour of sea lavender lends itself to a Mediterranean-style planting scheme or it can be used en masse in a contemporary garden.

Miscanthus sinensis 'Sarabande'
Japanese silver grass

Mature height and spread: 1–2 m **Spacing**: 1 plant per 1.2 m^2 **Plant form/habit**: Grass **Position**: Full sun **Leaf hold**: Deciduous **Benefits**: Flowers, foliage **Flowering time**: Mar–Aug **Flower colour**: Silver-pink turning red-purple then fading to pale straw **Climate**: Cold, cool temperate, warm temperate **Water requirement**: Moderate

Miscanthus is another of my favourite ornamental grasses. I often choose it over *Calamagrostis* × *acutiflora* 'Karl Foerster' (see page 222) for its height and volume. It tends to be bulkier and taller, adding substance to a border rather than finesse. The flowers or seed heads are a darker, more dramatic colour. This *Miscanthus* is completely dormant over winter, so the entire plant becomes an aged straw colour. The foliage, together with the feathery seed heads that are held above, persists to the end of winter.

The time to chop this grass to the ground is at the beginning of spring, when you start to see the fresh green growth poking through the ground. Every few years, the plant will need to be renewed by lifting and dividing clumps at the end of winter.

Miscanthus likes well-prepared soil, so add lots of compost and cultivate well. Do not over-water, though, as this will generate too much foliage and not enough flowers.

It grows tall, so place it at the very rear of the bed and clump three or four plants together.

Nepeta × faassenii
Catmint

Mature height: Up to 300 mm **Mature spread**: 300–500 mm **Spacing**: 6 plants per m^2 **Plant form/habit**: Clump-forming **Position**: Full sun **Leaf hold**: Evergreen **Benefits**: Flowers, foliage **Flowering time**: Oct–Apr **Flower colour**: Blue **Climate**: Cold, cool temperate, warm temperate **Water requirement**: Drought tolerant

Catmint is an essential plant for any country garden. I use it as the perfect underplanting for roses, as a single species for a long narrow bed, or simply in large clumps in flower borders. In summer it is covered in tiny blue flowers that last for months. Once flowering is finished, cut it back by 50 per cent and it will provide a second flush.

When positioning catmint in a border, place it at the front, as it is one of the lowest flowering perennials, forming a small clump of grey–green foliage typically to just 200 millimetres in height.

It likes a hot, dry summer and a cold winter, and will not tolerate any humidity, so its growth is limited to the southern states only.

Catmint is herbaceous, so cut it to the ground at the end of autumn. This plant is extremely tough, requiring very little water or maintenance.

Nepeta 'Dropmore'
Dropmore catmint

Mature height and spread: 300–500 mm **Spacing**: 2 plants per m^2 **Plant form/habit**: Clump-forming **Position**: Full sun **Leaf hold**: Deciduous **Benefits**: Flowers, foliage **Flowering time**: Nov–May **Flower colour**: Blue **Climate**: Cold, cool temperate, warm temperate **Water requirement**: Drought tolerant

'Dropmore' is a larger-growing catmint that is about double the height of *N.* × *faassenii* (see page 240). I tend to choose it over *N.* × *faassenii* as it flowers for twice as long: it is feasible to have it flower from late spring right through to the end of autumn. To encourage this, simply deadhead the spent flowers once during this period and it will maintain flowering.

I have used 'Dropmore' to great effect under a pergola of climbing roses and as an informal flowering hedge along a path.

It needs dry heat and well-drained soil. Cut it to the ground every winter and feed with compost annually.

Nicotiana langsdorffii
Flowering tobacco

Mature height: 500–1000 mm **Mature spread**: 300–500 mm **Spacing**: 1 plant per m^2 **Plant form/habit**: Upright **Position**: Full sun **Leaf hold**: Evergreen **Benefits**: Flowers, foliage **Flowering time**: Dec–Mar **Flower colour**: Lime-green **Climate**: Cold, cool temperate, warm temperate **Water requirement**: Moderate

The lime-green flowers of this delightful perennial hang like little funnels from tall flower stems and dangle and sway in the breeze. Although it is a perennial, I treat *Nicotiana* as an annual, pulling it out each winter and replacing it in the spring.

It grows easily from seed, which you can harvest and sow yourself. As the flowers come to their end, you will find the fruit capsules loaded with seed, so collect this in an envelope and sprinkle it back into the areas of the garden where you would like *Nicotiana* to re-emerge. The plant's flowers add vibrancy to the garden, and this warrants the trouble it takes to renew it each year.

Nicotiana should flower throughout the whole of summer and part of early autumn, with the foliage providing interest during spring. It gets tall and often leggy, so create a structure of branches around it for support as it is growing.

It likes some water in summer and low humidity, and prefers full sun. I use it only in flower borders, where I place it in clumps of three towards the rear of the bed.

Nierembergia caerulea (syn. Nierembergia hippomanica)
Cup flower

Mature height and spread: 300–500 mm **Spacing**: 4 plants per m^2 **Plant form/habit**: Rounded **Position**: Full sun **Leaf hold**: Deciduous **Benefit**: Flowers **Flowering time**: Nov–Mar **Flower colour**: Blue **Climate**: Cold, cool temperate, warm temperate **Water requirement**: Moderate

Nierembergia can be used en masse or even as a specimen in a flower garden, but I prefer to treat it as a perennial and plant it accordingly in my flowering herbaceous borders. I use it this way because it does better when treated as a herbaceous perennial; that is, cut down to the ground every autumn. It then responds with new growth in the spring and an extended flowering season from late spring to early autumn.
It can also be propagated by seed or division.

Nierembergia caerulea gets covered in small mid-blue flowers and grows to 600 millimetres, so plant it towards the front of the border, where it can best be appreciated.

Extreme cold during winter is not an issue for *Nierembergia*, and it loves hot, dry summers and a full-sun position. The plant will not tolerate humidity, though. A moderate amount of summer watering is required. Well-cultivated soils are important and it appreciates having compost and lime added to the soil annually.

Pennisetum alopecuroides
Foxtail grass

Mature height and spread: 500–1000 mm **Spacing**: 2 plants per m^2 **Plant form/habit**: Grass **Position**: Full sun **Leaf hold**: Deciduous **Benefits**: Flowers, foliage **Flowering time**: Jan–April **Flower colour**: Insignificant **Climate**: Cool temperate, warm temperate **Water requirement**: Drought tolerant

Foxtail grass is a native of Queensland and New South Wales and is one of the finer grasses I use. Slightly taller than *Poa labillardieri* (see page 249), it flowers more prolifically, thus providing a better show. It boasts a soft, elegant fountain of foliage with ornamental foxtail-shaped flowers from spring to autumn.

Like *Poa*, foxtail grass is tough and requires very little maintenance or water – though foxtail grass will also welcome wet soil. It prefers full sun and should be placed towards the centre of the border or used en masse by itself. It thrives in coastal areas as well as inland, making it very versatile. In winter, plants become dormant and untidy and should be cut to the ground.

Penstemon 'Willy's Purple'
Penstemon

Mature height and spread: 500–1000 mm **Spacing**: 4 plants per m^2 **Plant form/habit**: Upright **Position**: Full sun **Leaf hold**: Evergreen **Benefits**: Flowers, foliage **Flowering time**: Dec–Apr **Flower colour**: Cool purple **Climate**: Cool temperate, warm temperate **Water requirement**: Moderate

I have always been reluctant to use penstemons as I have never really had much luck with them. They have either been poor flowerers or a bit straggly in growth. I have since discovered 'Willy's Purple', though, which dispelled all the concerns I'd had previously.

This cultivar is a vertical, robust-growing plant with spires of the most intense purple flowers for half the stem – much like a foxglove. The flowers last up to three months, after which I cut the dead flower spikes off and am promptly rewarded with another flush of flowers. I have positioned a large clump of seven plants in the centre of my border, making it the purple centrepiece.

Penstemons like a moderate amount of water, so make sure you have a good supply if you want to include these in your garden. They also appreciate lots of compost and well-cultivated soil. Full sun is essential, otherwise the flowers become leggy and the plant is more susceptible to mildew.

Perovskia atriplicifolia
Russian sage

Mature height: 500–1000 mm **Mature spread**: 300–500 mm **Spacing**: 4 plants per m^2 **Plant form/habit**: Upright **Position**: Full sun **Leaf hold**: Deciduous **Benefits**: Flowers, foliage **Flowering time**: Jan–Apr **Flower colour**: Blue **Climate**: Cold, cool temperate, warm temperate **Water requirement**: Drought tolerant

I first discovered Russian sage in the Hamptons, where it is used extensively in mass planting, often next to large drifts of lavender. This silver-leafed plant with striking white stems emerges as fresh new growth every spring. Summer follows, with a mass of small blue flowers which reach approximately 1 metre in height. From a distance the flowers merge to form a blue haze. The blue gradually fades in the autumn and leaves finally fall for the winter. The white stems remain for visual interest until early spring, when they can be pruned back to the ground to make way for the new shoots.

Russian sage is very hardy, requiring very little water or maintenance. It forms a tight clump, which often takes some time to form, so don't be disappointed that it is not much good in the first year – it definitely gets better with age. If the plant keeps collapsing, simply bury it in the winter by adding 150 millimetres of soil to the top, covering it completely.

It requires full sun and a well-ventilated position. I use it towards the rear of the border, or as a mass planting on its own. Position it in large clumps of at least seven plants.

Poa labillardieri
Common tussock grass

Mature height and spread: 300–500 mm **Spacing**: 4 plants per m^2 **Plant form/habit**: Grass **Position**: Full sun to partial shade **Leaf hold**: Evergreen **Benefits**: Flowers, foliage **Flowering time**: Oct–Feb **Flower colour**: Blue-green **Climate**: Cold, cool temperate, warm temperate, subtropical **Water requirement**: Drought tolerant

The delicate seed heads of *Poa labillardieri* rise up above the foliage every year, but they are not prominent. The main substance comes from the foliage. This grass is not as showy and lush as the other grasses mentioned in this chapter.

I use tussock grass as far north as Brisbane and all the way down the coast, right through to Adelaide. It is a low grass, so place it towards the front of the border. I tend to use it on its own as a mass planting or combined with a limited palette of other plants such as lavender (see pages 104–9) or *Santolina* (see page 136).

This native grass is extremely hardy, doing well in sandy soils and salt wind. It doesn't need to be cut down to the ground every year but will benefit from doing so every two to three years. It likes full sun to partial shade and really does not require any water even during the summer.

Polygonatum × hybridum
Solomon's seal

Mature height: 500–1000 mm **Mature spread**: 300–500 mm
Spacing: 5–7 plants per m^2 **Plant form/habit**: Clump-forming
Position: Partial shade to full shade **Leaf hold**: Deciduous
Benefits: Flowers, foliage **Flowering time**: Nov–Dec
Flower colour: White **Climate**: Cold, cool temperate
Water requirement: Moderate

I first discovered the very decorative Solomon's seal in an old garden at Mt Macedon near Melbourne. It was the home of the late famed florist Kevin O'Neill and was a true mountain garden with deep rich soil, cool summers and plenty of water. In one area he had a mass of Solomon's seal growing that he often picked to sell in his shop. I was mesmerised by this plant – the way the flowers clung like tiny bells to the arching stems seemed so unusual yet so decorative. I have always remembered this vision and whenever an opportunity arises, I use this plant.

Although the flowers are only short-term, the foliage texture remains as a feature, with leaves sitting in two neat rows along an arching main stem.

Solomon's seal must be grown in a shaded, cool environment and receive plenty of water during summer. I have found that due to this plant's fleshy rhizomes it has been able to endure soil dryness, though this is not ideal. Not a plant to tolerate humidity, a mountain environment is really best. It is a herbaceous perennial, so it dies down in winter. It does not need staking and spreads easily throughout the garden.

Position Solomon's seal in clumps of at least five to seven plants in a woodland situation. It grows well under smaller trees such as maples and oaks.

Rhodanthemum 'African Eyes'
Moroccan daisy

Mature height: 300 mm **Mature spread:** 400 mm **Spacing:** 8 plants per m^2 **Plant form/habit:** Spreading **Position:** Full sun **Leaf hold:** Evergreen **Benefits:** Flowers, foliage **Flowering time:** Spring to autumn **Flower colour:** White with dark brown eye **Climate:** Cold, cool temperate, warm temperate **Water requirement:** Drought tolerant, but best to irrigate in extended periods of heat

I discovered this low-growing, white-flowering perennial when I was looking for an alternative to *Erigeron* (see page 223). *Rhodanthemum* has much better foliage, being grey and lush, and the plant doesn't require constant cutting to encourage flowers. It's a clump-forming plant, but will form large clumps that sit low to the ground and are perfect for planting under shrubs. They are planted under my *Hydrangea paniculata*, where they start flowering much earlier than the hydrangeas and continue to flower as the hydrangeas start. They love full, hot sun, and good drainage is a must. They tolerate frost well but not humidity. Don't cut them back too hard in winter, as this can kill the plant; simply reduce the foliage by approximately 50 per cent.

Salvia leucantha
Mexican bush sage

Mature height and spread: 1.2 m **Spacing**: 900 mm off-centre **Plant form/habit**: Rounded **Position**: Full sun **Leaf hold**: Evergreen **Benefit**: Flowers **Flowering time**: Feb–May **Flower colour**: Purple **Climate**: Cool temperate, warm temperate **Water requirement**: Low

Salvias are typically reliable and long-flowering plants. Some are harder to grow than others and may have varying flowering periods, but *S. leucantha* is one of the easiest and most prolific flowering salvias. In some warmer areas I have seen it flower all year round with the exception of a few cooler winter months. It will tolerate dry heat and some humidity. Its only drawback is that it can tend to flop when in full flower or when overwatered.

The deep purple flowers of *S. leucantha* are very dramatic in a herbaceous border and can dominate other perennials. It mixes well with shrub plantings and looks great with succulent plantings. It prefers full sun but will tolerate very light shade, and is drought hardy.

Salvia nemorosa
Woodland sage

Mature height and spread: 300–500 mm **Spacing**: 4 plants per m^2 **Plant form/habit**: Clump-forming **Position**: Full sun **Leaf hold**: Evergreen **Benefit**: Flowers **Flowering time**: Dec–Apr **Flower colour**: Purple **Climate**: Cold, cool temperate, warm temperate **Water requirement**: Moderate

This perennial is a small salvia, growing to only about 500 millimetres. It has the most intense purple flower and is covered in blooms from late spring, with deeply coloured bracts persisting into summer. Plant it at the front of the border and allow its colour to shine. Due to its low growth habit it is ideal to use as a break in a long border.

I recommend cutting back *S. nemorosa* to the ground after the first display of flowers to encourage a second floral burst in late summer to autumn. Like other salvias it prefers full sun, and unlike the bog sage (see page 255), it needs well-drained soil.

Salvia uliginosa
Bog sage

Mature height and spread: 1–2 m **Spacing**: 1 plant per m^2 **Plant form/habit**: Upright **Position**: Full sun to partial shade **Leaf hold**: Deciduous **Benefit**: Flowers **Flowering time**: Jan–May **Flower colour**: Blue **Climate**: Cold, cool temperate, warm temperate **Water requirement**: Moderate

There are so many salvias available, it can be hard to know which to choose. Bog sage is a perennial favourite of mine, and I have been using it throughout my career.

The deep blue flowers of this sage are carried on long stems and float in the breeze – they appear to glide above the garden bed. An incredibly tough plant, it will take some waterlogging of the soil, as its name implies.

It is a good idea to lift and divide this plant every year, digging over the soil to remove any runaway rhizomes and then replanting it in clumps where desired. This is a good method for keeping the plant young and in check.

Bog sage can spread very readily throughout the garden, so keep an eye on it and remove any wayward parts as they arise. It will grow in partial shade but much prefers full sun. Cut it down to the ground in winter.

Salvia 'Waverly'
Mexican bush sage

Mature height: 1.2 m **Mature spread:** 1.2 m **Spacing:** 1 plant per 1 m^2 **Plant form/habit:** Rounded **Position:** Full sun **Leaf hold:** Evergreen **Benefits:** Flowers **Flowering time:** Feb–Jun **Flower colour:** White with dark purple calyx **Climate:** Cool temperate, warm temperate, subtropical **Water requirement:** Drought tolerant, but best irrigated during extended heat

Salvias are one of the easiest and toughest perennials, and *S.* 'Waverly' is the toughest of them all. It flowers from October right through until May with white flowers that have a purple calyx and are attractive to birds. I use it in my blue flowering border. Its purple calyxes merge well with the blue hues and it doesn't appear as a pure white-flowering plant. If you want perennials in the more humid areas of Australia, this is one of the best ones to use. I have had it flowering well in Brisbane. The plant itself is a large clump of low, dark green foliage with large spikes of flowers. It is not my favourite, but I use it when I want reliable flowering or have a spot where nothing else will grow in the border. Plant in full sun, protect new growth from spring frosts, and cut it to the ground in winter.

Sedum spectabile × Sedum telephium (syn. Hylotelephium spectabile) 'Autumn Joy'

Sedum

Mature height and spread: 500–1000 mm **Spacing**: 2 plants per m^2 **Plant form/habit**: Rounded **Position**: Full sun to partial shade **Leaf hold**: Deciduous **Benefits**: Flowers, foliage **Flowering time**: Jan–Jun **Flower colour**: Green turning deep pink **Climate**: Cold, cool temperate, warm temperate **Water requirement**: Drought tolerant

One of the hardiest and most versatile plants I have used is the sedum. I find its form and hardiness utterly compulsive. At first I was attracted to 'Autumn Joy' for its green flower buds, which I simply love, as they are quite rare in the plant world. But after growing it I soon discovered how wonderfully tough it was. I was not only growing it in full sun but also under a dry and shaded cypress tree. It is now a key plant in my herbaceous border at my farm, Stonefields.

This plant is interesting all year round. Young growth erupts from the ground in early spring and quickly develops into a textured, tidy dome of fleshy leaves. During summer the stems start to extend and the umbelliferous flower heads develop. Summer to autumn sees the flower heads change colour from green, becoming flushed with pink and then increasingly darkening to a deep earthy pink in the autumn. Next they rust off to brown as the leaves begin to fall for the winter. The brown umbels are held right through wintertime and are highlighted white with winter dews and frosts. The cycle begins again in spring, and the old stems should be removed as the new growth pushes through once again. This is the right time to lift and divide if renewal is required.

Sedums are great as a massed planting or used in clumps within a flower border. However, they can be their own worst enemy as they flower so well they tend to fall apart due to the flowers' weight. I learnt in England that the trick is to cut them back hard about a month after they first emerge in the spring – this strengthens the growth and prevents them collapsing.

Sedums prefer a dry summer and don't do well when over-watered. They will tolerate some humidity, however.

Sedum telephium 'Matrona' (syn. Hylotelephium telephium)
Sedum

Mature height: 800 mm **Mature spread:** 600 mm **Spacing:** 5 plants per 1–2 m^2 **Plant form/habit:** Upright and clump forming **Position:** Full sun to partial shade **Leaf hold:** Deciduous **Benefits:** Flowers, foliage, stems **Flowering time:** Mar–May **Flower colour:** Pale pink, ageing to russet, then brown **Climate:** Cold, cool temperate, warm temperate **Water requirement:** Drought tolerant

For a long time I only used the well-known *Sedum* 'Autumn Joy', a reliable green-flowering sedum that turns terracotta in late autumn. Desiring more colour in my borders, I selected 'Matrona'; with its deep red stems and flowers it adds a sense of drama to any border's colour scheme. It's a plant that commands attention with its structure, even though it is subtle in foliage colour. Upright burgundy stems and large blue-green leaves with purple bloom combine well to foil the soft pink umbelliferous flowers, which are held atop the plant in late summer to early autumn.

The plant itself is also more sturdy, as the stems don't tend to fall apart. It will flower for up to four months of the year in summer, but its stems add colour, even when it's not in flower. It prefers a hot, dry environment, being susceptible to mildews in more humid conditions. Good drainage is a must. Cut it to the ground in winter and divide the clump every three years, or it will slowly weaken the robustness of the plant. I plant it to the front of my borders and usually in groups of three.

Stachys lanata
Lambs' ears

Mature height: Up to 300 mm **Mature spread**: 500–1000 mm **Spacing**: 4 plants per m^2 **Plant form/habit**: Groundcover **Position**: Full sun to partial shade **Leaf hold**: Evergreen **Benefits**: Flowers, foliage **Flowering time**: Nov–Mar **Flower colour**: Blue/purple **Climate**: Cool temperate, warm temperate **Water requirement**: Moderate

I believe *Stachys lanata* is the perfect plant that has a place in all country gardens. Grey foliage is often required to subdue a rather vibrant colour scheme, and this is the ideal grey foliage plant.

Stachys lanata sits low, which suits the front of a border, is incredibly tough, and provides colour through its foliage for most of the year. It has the added bonus of flowering with soft spikes of blue flowers. I don't plant it for this reason, but appreciate it when the flowers emerge.

The foliage of this plant is soft and can die back in frost or very cold weather. For this reason I tend to cut it down to the ground in late autumn. In Melbourne, Adelaide and Sydney it is capable of looking good for 12 months of the year.

Lambs' ears spread very fast and need containing to prevent them running into other plants. I do this with a sharp spade, merely cutting a circle around the plant to the size I require.

They love full hot sun and hate any humidity. Though capable of surviving in dry soils, lambs' ears present more attractive foliage when soil moisture is maintained. It is not fussy about soil types, but thrives in rich soils. Grow it simply by layering a cutting with a few roots on it in its desired location – it's that easy.

Trees

Above: *Malus floribunda*

Trees are a vital component of almost every garden. They provide scale, shade and mystery and, for children, they add magic.

I have an enduring childhood memory of my grandmother's enormous pine tree, which seemed enchanted. It sat at the bottom of the garden and my grandfather had turned it into the most amazing climbing tree by adding timber runs leading to a treehouse at the top. My memory of this tree is still strong, and is one of the reasons why I always try, where space permits, to include one significant tree in every garden, hoping that one day a child will get to enjoy it as I did my grandparents' pine.

Trees play a number of important roles in the garden. First, they serve an aesthetic purpose. They make gardens visually stimulating by providing them with a three-dimensional sense of space. This is vital because, without it, gardens can appear flat and boring. They provide depth when placed at the beginning of a vista and a focal point when placed at the end. And they help to punctuate the garden's design by providing vertical points of interest.

Trees also serve a vital functional purpose. They are a garden's main source of shade, allowing delicate, sun-shy woodland plants to grow, and helping to cool the garden and sometimes the house with their foliage.

My favourite type of tree is the oak (see pages 334–43). Oaks have a majestic, spreading shape; they are tough and resilient; and they have a sense of age and grandeur unequalled by other kinds of trees. The London plane tree (see page 326) is my next favourite, as it is the perfect avenue or driveway tree and can also be manipulated into a pleached hedge or umbrella-shaped tree. I also love lindens (see page 346) for their scented green flowers, large leaves and conical form that matures into the perfect specimen shape.

Trees, no doubt because they take a long time to reach maturity, are not subject to the whims of short-term gardening fashions. The types of trees used in gardens are, however, changing slowly as gardens become smaller. Most gardens today cannot afford

Above: *Tilia cordata*

the luxury of a liquidambar or plane tree. They are more likely to contain a small maple (see pages 268 and 274), redbud (see page 277) or ornamental pear tree (see pages 328 and 330). Trees must now be smaller and more decorative, and have less aggressive root systems. Gardeners demand more from their trees now than ever before in gardening history.

A final note of warning. Make sure that you fully understand the workings of your tree before selecting and planting, because trees can have disastrous impacts on gardens, other structures in the garden, and even on houses. I learnt this lesson the hard way early in my career. While searching for the ideal pleaching tree upon returning from my first European trip, I found *Populus yunnensis,* which was fast-growing, deciduous and responded well to trimming. I thought that I had found the perfect tree. It was not until a couple of years later that I discovered that its extremely aggressive root system was not suited to small inner-city gardens. Lesson learnt, I moved on and found ornamental pears, which have, in comparison, a very gentle root system.

Tips for selecting the right tree

1. Consider if you want year-round shade from an evergreen or winter sun to shine through with a deciduous tree.
2. Decide if a tree is for simple screening or if it is to provide a decorative purpose. This will help in your selection.
3. Consider your decision carefully before planting. Once grown, it is often difficult to gain council permission to remove a tree.
4. Check how extensive the root system of your tree will be when mature. Plant a safe distance away from buildings.

Acacia melanoxylon
Blackwood

Mature height: 10–20 m **Mature spread**: 4–10 m **Spacing**: Plant 10 m from other trees or large shrubs if you are planting a single tree; for a windbreak, plant them 3 m apart **Plant form/habit**: Rounded **Position**: Full sun to partial shade **Leaf hold**: Evergreen **Benefits**: Flowers, foliage **Flowering time:** Dec–Feb **Flower colour**: Cream **Climate**: Cold, cool temperate, warm temperate, subtropical **Water requirement**: Drought tolerant

If you favour the form of European trees but would prefer to plant a native tree, then the blackwood is for you. It is renowned for its attractive timber, which is used for furniture. It produces pale cream flowers but these are not really its main feature. Although wattles are often known for being short-lived, the blackwood is one of the longest living of this genus. I first saw the tree on an English park-like manor property at Mt Macedon in Victoria where they were scattered through a field. They looked as if Capability Brown himself had planted and pruned them.

Blackwoods are one of the toughest trees I use and tolerate a wide range of conditions. I often plant them, give them their initial water and then never water them again. Just plant and watch it grow!

The trick is to resist planting the blackwood as established stock and to plant small trees instead. If you use stakes and put rabbit guards around them they will grow into decent-sized trees within eight years. Just don't plant them too closely together as this will affect their mature shape.

Acer palmatum
Japanese maple

Mature height and spread: 4–10 m **Spacing**: Plant 2–3 m from other trees or large shrubs **Plant form/habit**: Vase **Position**: Partial to full shade **Leaf hold**: Deciduous **Benefit**: Foliage with yellow to orange-red autumn leaf colour **Climate**: Cold, cool temperate, warm temperate **Water requirement**: Moderate

The Japanese maple is one of my favourite and most used small trees. It is a manageable size for city gardens, fitting into courtyards or even smaller spaces. I have also used them successfully in large light wells. They come in a range of colours, from varieties that are green throughout the growing season turning to red, orange or yellow in the autumn, to others that are red from spring through to autumn. I often use them in copses, placing them randomly throughout beds or even in larger spaces throughout a lawn.

Unfortunately, this tree has suffered badly from the effects of climate change, and often burns in extreme summer sun. For this reason, in the drier southern states place Japanese maples in dappled light or full shade, though they should tolerate some morning sun. They do need water, however, so make sure you have an adequate supply before planting. Because of their vibrant colour range, and as they develop a highly sculptural trunk and branch structure, they are one of the most rewarding trees to uplight at night.

Japanese maples are ideal for underplanting with other smaller, more delicate plants as they have a non-aggressive root system. They look particularly good with star jasmine (see page 200) planted underneath.

Acer palmatum 'Atropurpurea'
Red-leafed Japanese maple

Mature height: 4–6 m **Mature spread:** 4–6 m **Spacing:** 6 m apart **Plant form/habit:** Vase, becoming rounded **Position:** Full sun to partial shade **Leaf hold:** Deciduous **Benefits:** Foliage (autumn) and branch form **Climate:** Cold, cool temperate, warm temperate **Water requirement:** Moderate

The deep purple shade of this maple is held throughout the growing season, making it a great dramatic feature in the garden. Most maples colour during the autumn and are often grown specifically for this dramatic effect, but this maple provides you with the effect all spring, summer and autumn. The dark colour to the leaf helps it tolerate summer sun and heat, although it can still burn in extreme summer conditions. Like all maples, they are excellent small trees with a non-aggressive root system, making them easy to grow other plants under, and for positioning close to a building.

Acer palmatum 'Senkaki'
Coral bark Japanese maple

Mature height: 4–6 m **Mature spread:** 4–6 m **Spacing:** 6 m apart **Plant form/habit:** Vase, becoming rounded **Position:** Full sun to partial shade **Leaf hold:** Deciduous **Benefits:** Foliage (autumn) and branch form **Climate:** Cold, cool temperate, warm temperate **Water requirement:** Moderate

I use this maple extensively, mainly for its dramatic red stems and branches, which contrast with its pale green foliage in the growing seasons and then harmonise with its orange/yellow autumnal leaf colour. It tends to burn in the summer sun for its first few years of growth but will harden to more summer sun with age. It's an excellent small tree that can be used as a specimen or in closely grouped copses. I also often use it as a potted plant with great success. As with all maples, possums love this tree, so plant it only if possums are not a problem or if you can possum-proof the tree. Also referred to as *Acer* 'Sango Kaku'.

Acer truncatum × Acer platanoides 'Warrenred'
Pacific sunset maple

Mature height and spread: 4–10 m **Spacing**: Plant 10–12 m from other trees or large shrubs **Plant form/habit**: Rounded **Position**: Full sun **Leaf hold**: Deciduous **Benefit**: Foliage with red autumn leaf colour **Climate**: Cold, cool temperate, warm temperate **Water requirement**: Moderate

Pacific sunset is similar to the larger-growing sugar maples (*Acer saccharum*) and has good-sized large leaves with texture. I use Pacific sunset for its amazing, intense red colouring in the autumn. It also makes for a quick-growing, handsome tree for summer shade and form.

It can be hard to get other plants to grow under the Pacific sunset maple tree due to its shallow and greedy extensive root system, so plant it in an area where the view under the tree is not vital.

Pacific sunset can burn in extreme summer heat and is prone to leaf drop if it dries out in summer. For this reason it needs water, so make sure you have a source during the summer months. This isn't a large tree, making it useful for mid-sized city gardens or small country gardens.

Caesalpinia ferrea
Leopard tree

Mature height and spread: 10–20 m **Spacing**: Plant 15 m from other trees or large shrubs **Plant form/habit**: Rounded **Position**: Full sun to partial shade **Leaf hold**: Deciduous **Benefits**: Flowers, foliage, ornamental bark **Flowering time**: Sep–Nov **Flower colour**: Yellow **Climate**: Warm temperate, subtropical **Water requirement**: Moderate

It can be hard to find small specimen trees suitable for subtropical or humid areas of Australia but this is one such tree. I love its smooth, decorative bark with ornamental cream, brown and salmon patches, which are more pronounced than those of the crepe myrtle. The small leaflets are bipinnate – having feathery-looking compound leaves that grow opposite each other on a large stem – and provide a visually soft canopy. I like it for its graceful form, where the foliage seems almost to float on top of the tree, leaving the highly sculptural qualities of the trunk to be admired.

This tree will tolerate periods of dryness; however, it really only does well in humid situations, so don't try growing it in the southern states. It is deciduous, has small yellow flowers, likes well-drained, deep soil and tolerates some frost.

Cercis canadensis 'Forest Pansy'
Redbud

Mature height and spread: 4–10 m **Spacing**: Plant 4–5 m from other trees or large shrubs **Plant form/habit**: Spreading **Position**: Full sun to partial shade **Leaf hold**: Deciduous **Benefits**: Flowers, foliage with burgundy autumn leaf colour **Flowering time:** Sep–Nov **Flower colour:** Burgundy **Climate**: Cold, cool temperate, warm temperate **Water requirement**: Moderate

With its deep plum-coloured summer foliage I love this small tree as it creates much drama in the garden. The redbud isn't tall and for this reason it works well in courtyard situations. Its foliage, not being dense, allows plenty of light underneath for lower-storey plants to thrive.

Recently I've seen redbuds burn in the hot summer sun, so if you live in the southern states, choose a site protected from the afternoon rays. They will cope with humidity in more northern areas and tolerate the afternoon sun better there. Plant them with plenty of space around them, in order to enjoy their whole canopy.

Cornus florida
Flowering dogwood

Mature height and spread: 4–10 m **Spacing**: Plant 3 m from other trees or large shrubs **Plant form/habit**: Vase **Position**: Full sun to partial shade **Leaf hold**: Deciduous **Benefits**: Flowers, foliage with autumn leaf colour **Flowering time**: Oct–Nov **Flower colour**: White, pink **Climate**: Cold, cool temperate **Water requirement**: Moderate

Dogwoods are one of those trees that many of us cannot grow because they require really cold conditions, but are so special that nearly everyone wants to try! To see this tree covered in spring flowers that appear to float on the bare-looking branches is a magical sight. They come in pink and white, but I prefer the pure whites, with 'Eddie's White Wonder' being one of my favourites.

Flowering dogwoods really love a cold winter and relatively cool summer, as they can certainly burn in our Australian heat and sun. If you are fortunate enough to be in an elevated area or a cool state, then they should grow successfully. I am fortunate to be able to grow them at my country property, Stonefields, where I have planted them encircling the round lawn in the woodland. The best time to view them in flower is under a full moon where they take on a luminescent glow.

They are small, delicate trees and can be very spectacular – so if you want to create drama and show in your garden, you won't be disappointed.

Cupressus sempervirens 'Glauca'
Pencil pine

Mature height: 10–20 m **Mature spread**: 1–2 m **Spacing**: Plant 3 m from other trees or large shrubs **Plant form/habit**: Upright **Position**: Full sun **Leaf hold**: Evergreen **Benefit**: Foliage **Climate**: Cool temperate, warm temperate **Water requirement**: Drought tolerant

The pencil pine is another tree that evokes the feeling of a Tuscan or Mediterranean garden. One of the most beautiful drives I have ever seen was an avenue of 300-year-old pencil pines in Tuscany. In its old age the tree develops a clean trunk to about chest height, which then gives way to a tight column of dark green foliage. Commanding attention in the garden, these trees make the perfect accent tree with their almost exclamation-point form. I use them to mark corners or as a central feature in the centre of a garden space.

Unfortunately, being a cypress, they are prone to canker, and I have seen a lot succumb to this disease. There seems little you can do about it, except for cutting out the infected wood and hoping the tree outgrows it. Copper sprays may provide a level of protection.

There are two closely related types of pencil pine: 'Glauca' and 'Stricta'. I use the former as it seems to hold its foliage tighter and not fall apart as 'Stricta' can often do.

When staying in Italy once I saw an old Italian gardener pruning a young plant. When I asked him what he was doing, he replied that all pencil pines in Italy are trimmed as young plants to help shape them into that tight form that we often associate with the plant. I have since trimmed all my pencil pines by gently clipping their sides twice a year while they are growing.

Delonix regia
Poinciana tree

Mature height and spread: 4–10 m **Spacing**: Plant at least 20 m from other trees or large shrubs **Plant form/habit**: Spreading **Position**: Full sun **Leaf hold**: Deciduous **Benefits**: Flowers, foliage **Flowering time**: Dec–Feb **Flower colour**: Orange-red **Climate**: Subtropical, tropical **Water requirement**: Moderate

The poinciana is a very ornamental tree for its luxuriant foliage alone. The flowers are a rich deep orange-red and are abundant over the summer. The tree is highly sculptural and has the unique form of growing almost perfectly laterally, thus creating an umbrella effect with its branches. It has wonderful soft, highly dissected leaves that allow dappled light through. Due to its vast spreading nature I use it as a shade tree in the northern parts of Australia, where shade is often much needed in the garden. Unfortunately this is not a tree for the southern states, and it does not tolerate frost.

The poinciana tree does become large, not so much in terms of height, but due to its spreading nature. The root system can become a problem as it has shallow major roots that often break the soil surface. The roots can become so matted that it is hard to grow other plants under it. It has large seed pods that are very decorative but do drop and cause some mess, so be aware if placing it near a swimming pool.

Diospyros kaki
Persimmon

Mature height and spread: 4–10 m **Spacing**: Plant 5 m from other trees or large shrubs **Plant form/habit**: Rounded **Position**: Full sun **Leaf hold**: Deciduous **Benefits**: Foliage with red autumn leaf colour, fruit **Climate**: Cold, cool temperate, warm temperate, subtropical **Water requirement**: Moderate

Many fruit trees are both productive and ornamental and the persimmon is one of my favourites in this group. This tree has an attractive deep glossy leaf that turns a majestic red in autumn, after which the orange fruit hang onto their bare branches during the winter period. This makes them highly sculptural, and as they are only a small tree, they suit a modern inner-city yard as well as groves in country gardens. They tolerate a wide range of conditions, from cold winters to hot dry summers. They love full sun and should be placed in a spot where they can be viewed on all sides. They form a natural umbrella shape, thus lending themselves to uplighting.

Dracaena draco
Dragon tree

Mature height and spread: 4.5–7 m **Spacing**: Feature as per plan **Plant form/habit**: Spreading **Position**: Full sun, partial shade **Leaf hold**: Evergreen **Benefit**: Feature tree **Flowering time**: Nov–Dec **Flower colour**: Green/white **Climate**: Cool temperate, warm temperate, subtropical **Water requirement**: Drought tolerant

The dragon tree is a small subtropical tree that will tolerate humidity as well as cooler climates. It is extremely slow-growing and will live for hundreds of years. It is very expensive when bought as a mature plant and is therefore used sparingly in a garden. It grows in full sun or shade and can even be used as an indoor plant.

The dragon tree is a very strong architectural plant, so place it where you want to draw attention to a certain part of the garden. It makes an excellent accent tree for spaces where a dramatic statement is needed.

Ficus carica
Fruiting fig

Mature height and spread: 2–4 m **Spacing**: Plant 4 m from other trees or large shrubs **Plant form/habit**: Rounded **Position**: Full sun **Leaf hold**: Deciduous **Benefits**: Foliage, fruit **Climate**: Cool temperate, warm temperate **Water requirement**: Drought tolerant

I have included this fig in the Hedges chapter (see page 37, where there is lots of information about this tree), as it is one of my very favourite trees for hedging, as well as here and in the Textural Plants section (see page 363). So you can see I really do love this tree and use it in a variety of ways!

The fruiting fig does make a very strong, sculptural specimen tree with the added benefit of fruit. You can leave this tree to grow into its natural height and form and it will still fruit prolifically. I prefer to shape it slightly, however, creating a round ball contour to about 4 metres in height.

It's one of those rare trees that is productive as well as having a magnificent form. Figs evoke a strong sense of history – of antiquity. Fig leaves, along with acanthus leaves, are often portrayed in ancient Greek and Roman sculptures.

Fraxinus spp.
Ash

Mature height and spread: 10–20 m **Spacing**: Plant 15 m from other trees or large shrubs **Plant form/habit**: Rounded **Position**: Full sun **Leaf hold**: Deciduous **Benefit**: Foliage with red or yellow autumn leaf colour **Climate**: Cold, cool temperate, warm temperate **Water requirement**: Drought tolerant

Ash trees are inconspicuous in the garden for most of the year until the cooler days of autumn, when these trees truly come into their own. Then they have a good show of colour that holds well when compared to other autumn trees. They definitely signal that summer has come to its end. Exhibiting a longer period of dormancy than most deciduous trees, they are often the first to reveal autumn colour and the last to burst into bud in the spring. I like the golden ash (*F. excelsior* 'Aurea'), which turns yellow in autumn, and claret ash (*F. angustifolia* 'Raywood'), which displays a deep claret red.

The ash isn't a graceful tree in my view, but rather robust and strong in nature. Its trunk is one of its main features – as it matures it starts to become very textured and rough and usually very straight. Ash trees are known for their ability to survive through dry conditions; in fact they prefer it dry instead of too wet.

I don't recommend them often, mainly because their growth habit is slow, but when clients are patient, or have the necessary space in a rural situation, I love to use them. The foliage is never dense, casting a dappled light to the bed underneath and allowing you to grow a good diversity of plants. For this reason also they are a very suitable tree for growing in a lawn as the dappled shade they produce allows the grass to grow under them.

I use them sparingly in my landscapes, placing only one or two in a large city garden and a limited number in country gardens.

Right: *Fraxinus excelsior* 'Aurea'

Ginkgo biloba
Ginkgo, maidenhair tree

Mature height: 20+ m **Mature spread**: 10–20 m **Spacing**: Plant 6–7 m from other trees or large shrubs **Plant form/habit**: Upright **Position**: Full sun **Leaf hold**: Deciduous **Benefit**: Foliage with yellow autumn leaf colour **Climate**: Cold, cool temperate, warm temperate **Water requirement**: Drought tolerant

The ginkgo is one of the most ancient plants. It is known to have been in existence for tens of thousands of years. It has the most amazing show in the autumn of intense butter-yellow foliage. These trees get better with age, not really reaching their best until they are at least 10 years old. Tall and narrow, with a very rigid branching framework, the leaves are a very distinct lime green, with a shape reminiscent of a Chinese fan. They make a good feature tree in the garden, being conical in shape. They are extremely tough – proven by being one of the main street trees in Manhattan, growing out of concrete in the hot and cold extremes of New York!

These trees are dioecious, meaning they bear male and female flowers on separate trees. Select male pollen-bearing trees to grow, because the mature female trees bear fruit that has an unpleasant odour. Possums love them, so keep this tree from boundaries and fences, which seem to be possum highways. Ginkgos can grow in nearly all states of Australia, but they do prefer a cold winter and not too much humidity.

Jacaranda mimosifolia
Jacaranda

Mature height: 10–20 m **Mature spread**: 10–20 m **Spacing**: Plant 10–12 m from other trees or large shrubs **Plant form/habit**: Irregular **Position**: Full sun **Leaf hold**: Deciduous **Benefits**: Flowers, foliage **Flowering time:** Nov–Dec **Flower colour:** Cool purple **Climate**: Warm temperate, subtropical **Water requirement**: Moderate

The very popular jacaranda is a glorious seasonal tree with an irregular branching structure. It comes into full bloom in early November in Sydney and December in Melbourne, when it is smothered in long-lasting cool purple flowers, and heralds that Christmas is near. The flower colour can vary from soft blue through to purple. When the trumpet-shaped flowers drop, they form a beautiful purple carpet underneath the tree. After the flowers fall the soft, lush and bright grass-green foliage of the jacaranda tree begins to develop, each leaf divided into tiny bipinnate leaflets (compound leaves that grow opposite each other on a large stem). These lacy leaves are held right through the winter, slowly yellowing and then falling to reveal the branches in the spring. This natural shedding period occurs twice a year, and bear in mind it can produce a lot of leaf and blossom litter.

Jacarandas grow well in most areas of Australia; however, they aren't suited to Tasmania or cold mountainous regions. They prefer a sunny position; rich, well-drained soil; and protection from wind and frost when young. Try to avoid pruning; only remove broken or diseased branches, otherwise the tree can send up vertical shoots that can distort its shape. The jacaranda is disease- and pest-resistant.

Lagerstroemia indica 'Natchez'
Crepe myrtle

Mature height and spread: 4–10 m **Spacing**: Plant 5–7 m from other trees or large shrubs **Plant form/habit**: Vase **Position**: Full sun **Leaf hold**: Deciduous **Benefits**: Flowers, foliage with autumn leaf colour **Flowering time**: Feb–Mar **Flower colour**: White **Climate**: Cool temperate, warm temperate, subtropical **Water requirement**: Drought tolerant

When searching for the perfect drought- and heat-tolerant small tree you could not look past the crepe myrtle, and my preference is the white blooms of the mildew-resistant variety 'Natchez'.

I had seen the crepe myrtle growing in Australia for many years, often coming across them in old gardens where they had been cut back to the same point every year, thus forming a small pollarded tree. But I really first became attracted to them when I saw a stunning crepe myrtle hedge at the new Getty Museum in Los Angeles. Here they were growing as a perfect drought-tolerant hedge in full white bloom. Other colours available include soft pinks, deep pinks, mauves and reds.

Crepe myrtles have the most interesting trunk in terms of colour variation, shifting from greys to light browns. The tree can be trained as a single trunk for pleaching purposes or encouraged as a multi-branched specimen to exhibit more of its smooth, stylish ornamental bark. They tolerate hot dry conditions to humid summers, growing equally well from Perth to Brisbane. They are not a large tree; in fact they are relatively small, so can easily be used in a courtyard situation. They are deciduous and I believe look good in all seasons: in flower, with autumn colour and as a skeleton in winter. Even better, they are pest- and disease-free.

Leptospermum laevigatum
Coastal tea tree

Mature height and spread: 2–4 m **Spacing**: Plant 5 m from other trees or large shrubs or every 1 m for a tight copse **Plant form/habit**: Irregular **Position**: Full sun to partial shade **Leaf hold**: Evergreen **Benefits**: Flowers, foliage with autumn leaf colour **Flowering time:** Dec–Feb **Flower colour:** White **Climate**: Cool temperate, warm temperate **Water requirement**: Drought tolerant

Another decorative and versatile native tree I love is the tea tree, which leans with the direction of the wind. This tree you see almost always growing diagonally, which is how it should be, with trunks even more beautiful and gnarled than the moonah (see page 313). I grow it as a multi-stemmed tree; that is, it doesn't have one main trunk but many. This adds to its sculptural quality.

I associate the tea tree with coastal regions and therefore only use it in beachside gardens. Here it thrives in the sandy soils and loves the exposure to wind and sun. It's extremely sculptural, so position it on a lawn or surround it with low planting where its beauty can be appreciated. This tree is all about the trunk and branches, with the leaves being of little interest. For this reason it is a perfect plant to uplight.

It loves the full sun but will also survive in semi-shaded areas. It is not a tall tree, so can therefore be planted in large groups. Its root system is not invasive so it is an ideal tree to plant close to a house. I have only seen this plant growing on the mid-to-southern east coast of Australia and for this reason I have limited its use to these areas.

Liquidambar styraciflua
Liquidambar

Mature height: 20+ m **Mature spread**: 10–20 m **Spacing**: Plant 12 m from other trees or large shrubs **Plant form/habit**: Pyramidal **Position**: Full sun **Leaf hold**: Deciduous **Benefit**: Foliage with autumn leaf colour **Climate**: Cold, cool temperate, warm temperate **Water requirement**: Moderate

For true brilliant autumnal colour, liquidambars cannot be beaten. Ranging from deep vibrant reds to rich butter yellows, they tend to hang on to their colouring for a good period of time. They are pyramid shaped, which means they do not take up too much lateral space in the garden, and for this reason they are often planted in reasonably small gardens. Spiky fruit drops to the ground, however, and should be removed as it can be difficult to walk on.

Liquidambar grows well in most states but requires some supplementary water during the summer months. They are reasonably fast growing but with this rapid growth comes one of this tree's major faults: its shallow root system. They often have a lot of major surface roots sitting above ground level. This makes planting underneath difficult and paving on top of them almost impossible. I would not like to plant them too close to a house as they may cause damage to the building's foundations. I prefer to limit them to country gardens or large city gardens. I use them as solitary trees, placing them as a major focal point within the garden.

Liriodendron tulipifera
Tulip tree

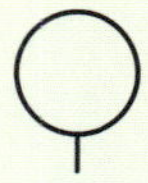

Mature height: 20+ m **Mature spread**: 10–20 m **Spacing**: Plant 15 m from other trees or large shrubs **Plant form/habit**: Rounded **Position**: Full sun **Leaf hold**: Deciduous **Benefits**: Flowers, foliage with autumn leaf colour **Flowering time:** Nov–Dec **Flower colour:** Green and orange **Climate**: Cold, cool temperate, warm temperate, subtropical **Water requirement**: Moderate

The tulip tree is one of the few large flowering deciduous trees of any note, and its flowers are one of its main attractions. As the name suggests, the unusual flowers resemble a large green and orange tulip. They do not typically appear on the tree for 8–13 years of age, however, so do not be too impatient to see them.

The tree itself is a tall, narrow form, making it ideal for medium- to large-sized urban gardens. It turns a wonderful buttery hue in the autumn, and tends to hold its colour for a good length of time. There is an even more upright form of this tree available as *Liriodendron tulipifera* 'Fastigiata'.

The tulip tree can be a little delicate in its need for water and intolerance of extreme heat. I have seen it grow well in Brisbane, Sydney, Melbourne and Tasmania, but I suspect the temperatures are getting too hot in Perth and possibly Adelaide now. It does love a cold autumn and winter, which helps create its rich autumn colouring. It holds its foliage well in the cold areas but less well in warmer climate zones. I would only use this tree as a single specimen or as a pair in a garden, as they are too special to plant en masse.

Magnolia grandiflora
Bull bay magnolia

Mature height and spread: 10–20 m **Spacing**: Plant 3–4 m apart to form a loose pleached hedge **Plant form/habit**: Rounded **Position**: Full sun to partial shade **Leaf hold**: Evergreen **Benefits**: Flowers, foliage, fragrance **Flowering time**: Jan–May **Flower colour**: White **Climate**: Cold, cool temperate, warm temperate, subtropical **Water requirement**: Moderate

Sometimes my clients want an evergreen tree in their garden, and as many of the more suitable inner-city trees are deciduous, I often suggest *Magnolia grandiflora* in these cases. Its rewards are many: it can be shaped and pruned to any form and size, it has magnificently scented white saucer-shaped flowers, green glossy leaves and, of course, is evergreen. It tolerates heat, cold and both dry and humid conditions, and it is one of those rare trees that will grow in every state of Australia. 'Exmouth' and 'St Mary' are my two favourite varieties, as I find they hold their leaf well and flower prolifically.

The leaf has a brown fur to the underside, which I love. Some don't agree, however, and for them there is now a form called 'Greenback' which has no brown colouring.

Some important factors for the success of evergreen magnolias include the need for a well-drained soil, regular soil moisture and fertiliser. These trees thrive in deep and moist humus-rich soil. Other than this, the species is fortunately low-maintenance and resistant to pests and disease.

Generally, this magnolia will tolerate a cold winter but I have seen it burn in a black frost. In England and America they often espalier this tree on the sunny side of the house or wall to protect it from the snow or frost. It is particularly beautiful when used this way, and I have copied this use extensively, not for climatic reasons but more on aesthetic grounds.

Magnolia grandiflora will grow in full sun or partial shade; if grown in full shade it tends to become a bit 'leggy', however. Do not be afraid to prune it – the Italians clip this tree into a cone shape which helps it to flower and create dense foliage. I carry out regular tip-pruning as the trees are growing too, as opposed to several heavy prunes.

Magnolia grandiflora Teddy Bear®
Southern magnolia

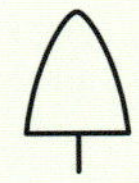

Mature height: 5 m **Mature spread:** 3 m **Spacing:** As specimen or 1 m apart as hedge **Plant form/habit:** Pyramidal **Position:** Full sun **Leaf hold:** Evergreen **Benefits:** Flowers and foliage **Flowering time:** Spring to autumn **Flower colour:** Creamy white **Climate:** Cold, cool temperate **Water requirement:** Moderate

The dwarf forms of *Magnolia grandiflora* have been around for some time now, the most popular being Little Gem. I have found the earlier ones to be prone to defoliating after five years of growth, causing the tree to look sick and very unattractive. The new form is Teddy Bear, and it seems to be performing much better. The appearance is denser in foliage with a more compact appearance. It grows slowly into a very tight and dense, pyramid-shaped small tree if left to its natural habit. It has glossy, thick, luxuriant, large, dark green leaves with fuzzy 'teddy bear'-brown reverse, and sporadic, large, fragrant, creamy white, saucer-like flowers from spring to autumn.

I have used it extensively as a clipped hedge and pleached hedge. When using it as a hedge, be aware that its root system can be very mat-like, making the growing of tender plants difficult underneath it. I have had success with *Murraya panniculata* and *Trachelospermum jasminoides* as an under-planting. The more you clip it the denser it will grow. Like all magnolias, it likes lots of water and fertiliser during the growing season. Its other distinct advantage is that possums seem to eat it last in the garden and will often not touch it at all. Its evergreen nature and compact growth habit make it an excellent tree for small gardens.

Magnolia × soulangeana
Saucer magnolia

Mature height and spread: 4–10 m **Spacing**: Plant 10 m from other trees or large shrubs **Plant form/habit**: Irregular **Position**: Full sun to partial shade **Leaf hold**: Deciduous **Benefits**: Flowers, foliage **Flowering time**: Aug–Sep **Flower colour**: Pale pink, white **Climate**: Cold, cool temperate, warm temperate **Water requirement**: Moderate

One of the most amazing visions I have ever seen is of two large *Magnolia* × *soulangeana* in full bloom in late snow outside the Frick Museum in New York. This tree is deciduous and flowers in spectacular fashion in late winter, when its branches are devoid of any foliage.

Unlike its hardier evergreen cousin, this magnolia has suffered badly in our extreme heat. Where it once made it through the summer unscathed, in the southern states this tree now seems to burn every summer. It is especially prone to burning early in the season with its new growth, so beware in the months of November and December. Give it protection from hot winds and shelter from the afternoon sun and it is still possible to grow it successfully.

Possums love the buds of this tree, so make sure it can be located away from possum highways such as power lines, other trees and walls. It is a spreading tree and matures into a very sculptural growth, so resist pruning it and allow its natural form to develop. It does like some humidity, so it does well in Sydney and not badly in Brisbane. It needs water in the summer, so make sure you have a good supply. I use it as a specimen tree and rarely as a grouping. Often I plant it in a lawn but more usually in a bed as its roots allow plants to thrive under it. This magnolia is pest- and disease-free.

Magnolia stellata
Star magnolia

Mature height and spread: 2–4 m **Spacing**: Plant 2–3 m from other trees or large shrubs **Plant form/habit**: Rounded **Position**: Full sun to partial shade **Leaf hold**: Deciduous **Benefit**: Flowers **Flowering time**: Aug–Sep **Flower colour**: White **Climate**: Cold, cool temperate, warm temperate **Water requirement**: Moderate

This is the smallest of the magnolias, growing into a compact tree. It also flowers on bare branches with smaller star-shaped flowers that are more prolific than the saucer magnolia (see opposite). I use it more as a large shrub at the rear of planting beds and often in groups of three or more. It also burns in extreme heat, so plant it in a position where it is protected from northerly winds and the afternoon sun. It will tolerate some humidity, but prefers cool winters and mild summers. The spectacular show of white flowers in late winter and early spring makes it a valuable addition to any garden bed. Again, this magnolia is pest- and disease-free.

Malus x domestica Crimson Crisp™ (cultivar 'Co-op 39') PBR
Edible apple

Mature height: 4 m **Mature spread:** 4 m **Spacing:** 6–8 m **Plant form/habit:** Broadly rounded, but can be formally trained **Position:** Full sun **Leaf hold:** Deciduous **Benefits:** Flowers, fruit **Flowering time:** Spring **Flower colour:** White, blushed with pink **Climate:** Cold, cool temperate, warm temperate **Water requirement:** Drought tolerant, best to irrigate while it's establishing itself

Most people associate fruiting apples with orchards and more productive areas of the garden, but for some time now I've been using them as a decorative tree that has the added benefit of being productive. I use them more often as an allée tree – creating an apple walk is a wonderfully romantic element to any garden. Their relatively small size allows you to create an avenue or allée on a very domestic scale. Once pruned, they provide the garden with sculptural form in the winter, soft abundant blooms in the spring and colourful fruit in the late summer and autumn. They tolerate a range of climates and soil types. Their root system allows for a variety of underplanting, from flowering groundcovers to clouds of *Buxus* balls. This variety is selected for its resistance to apple scab disease, and can be cross-pollinated with Granny Smith, Jonathon, Pixie Crunch™, and also Ballerina® cultivars with matching flowering times.

Malus floribunda
Japanese crabapple

Mature height and spread: 4–10 m **Spacing**: Plant 4–5 m from other trees or large shrubs **Plant form/habit**: Rounded **Position**: Full sun to partial shade **Leaf hold**: Deciduous **Benefits**: Flowers, foliage with autumn leaf colour, fruit **Flowering time**: Sep–Oct **Flower colour**: Pink turning white **Climate**: Cold, cool temperate, warm temperate **Water requirement**: Moderate

I first saw this crabapple when working in the Hamptons, where it is used extensively as a feature tree in lawns. To see it in spring is a magnificent sight – the tree is smothered in warm pink flowers that fade to white as they open fully. They cover the entire branch structure, producing an abundance of blooms reminding me of fairy floss. For this reason, crabapples are one of my favourite groups of compact trees. They are a manageable size, making them great for small gardens, and in addition to their spectacular show of flowers, they have beautiful autumn foliage colour, and some provide small fruit in the autumn.

I have planted Japanese crabapples in a field at my farm, using them in groups of three or five to soften the corners of the formal gardens as they protrude out into the softer landscape of the home field. They work well in groups and as with most trees these groups should always be made up of odd numbers and positioned 4–5 metres apart.

I find this tree very difficult to prune without ruining its natural shape. It tends to send branches softly arching in all different directions, and I think you just need to let it govern its own shape by simply pruning to remove crossing branches in order to prevent bark rubbing or stem damage. I also trim them so they have a clear trunk to approximately 1.2 metres. This way, I find it naturally develops into a rounded canopy.

Malus ioensis 'Plena'
Bechtel crabapple

Mature height and spread: 4–10 m **Spacing**: Plant 3 m from other trees or large shrubs **Plant form/habit**: Rounded **Position**: Full sun **Leaf hold**: Deciduous **Benefits**: Flowers, foliage with autumn leaf colour, fruit (inedible) **Flowering time**: Oct–Nov **Flower colour**: Pale pink **Climate**: Cold, cool temperate, warm temperate **Water requirement**: Drought tolerant

This crabapple is distinct due to its round shape: it forms an almost perfect ball sphere, as high as it is wide. In spring it is covered with a mass of pale pink flowers that are held well, unlike other crabapples that are prone to losing them in the wind. These good-sized double flowers are much larger than those of other crabapples. This tree also has marble-sized pale pink buds, which are beautiful even before the flowers open. Its foliage is denser than most and is pale green in colour, turning yellow in the autumn.

Like all other crabapples, it prefers a cold winter and warm summer, and the higher the humidity the less it flowers. Plant it in full sun or where it will receive late afternoon shade.

It does respond well to pruning: I have created a beautiful flowering pleached hedge from this plant. The drawback with this crabapple is its thorns, which can be dangerous to children.

Malus sargentii
Sargent's crabapple

Mature height: 1–2 m **Mature spread**: 2–4 m **Spacing**: Plant 6 m from other trees or large shrubs **Plant form/habit**: Spreading **Position**: Full sun **Leaf hold**: Deciduous **Benefits**: Flowers, fruit (inedible) **Flowering time**: Sep–Oct **Flower colour**: Pale pink turning white **Climate**: Cold, cool temperate, warm temperate **Water requirement**: Drought tolerant

This crabapple is a moderately fast-growing tree that is very hardy and flowers prolifically. It has the smallest of all crabapple leaves and flowers, making it very delicate-looking. It flowers in early spring with pale pink flowers fading to white.

Sargent's crabapple is wonderfully hardy, tolerating sun and heat well, though it does not like humidity. It requires little maintenance or water. I trim it so that it has a tall, clear trunk of say 1.5 metres, so that it keeps its canopy held high.

Malus spectabilis
Chinese crabapple

Mature height and spread: 2–4 m **Spacing**: Plant 4 m from other trees or large shrubs; plant 3–4 apart for an avenue **Plant form/habit**: Upright **Position**: Full sun **Leaf hold**: Deciduous **Benefits**: Flowers, foliage, fruit **Flowering time**: Oct–Nov **Flower colour**: Pale pink turning white **Climate**: Cold, cool temperate, warm temperate **Water requirement**: Moderate

This crabapple has sparse foliage on long, upright, slightly arching branches and produces an abundance of pale pink flowers fading to white in the spring. Its fruit is yellow and appears in the autumn, and as the fruit is not prolific the tree is often favoured for this reason.

Its vertical nature makes it ideal for creating a small avenue or walk. I used it successfully this way for my pond garden at St Ambrose farm. Alternatively it could be used as a specimen featuring in a lawn or gardenbed.

It is a reliable crabapple, but does require full sun to flower properly. Like all crabapples it enjoys a cold winter and dry hot summer. Low humidity is desirable and some water in summer is necessary.

Malus trilobata
Upright crabapple

Mature height: 4–10 m **Mature spread**: 2–4 m **Spacing**: Plant 4 m from other trees or large shrubs; plant 2 m apart for an avenue **Plant form/habit**: Upright **Position**: Full sun **Leaf hold**: Deciduous **Benefits**: Flowers, foliage **Flowering time**: Nov–Dec **Flower colour**: Pale pink turning white **Climate**: Cold, cool temperate, warm temperate **Water requirement**: Moderate

Malus trilobata is a tight-growing crabapple, slightly sculptural in its form, with deeply dissected maple-like leaves. Its very upright character makes it suitable for creating accents within the garden – think of it as a flowering pencil pine and use it in the same fashion. Its leaf colour is sage green with a slight grey tint, making it suitable for a grey foliage garden. It does not flower as spectacularly as the other crabapples, but still has a worthwhile pale pink to white flower.

Due to its heavily dissected leaf I use it chiefly to add texture to the garden. In fact, the leaf is almost more important to me than the flower.

It is a great tree to use when space is limited as it has a very limited spread. Like all crabapples it needs full sun and some water in the summer. It produces no real fruit to speak of. Its main feature is its vibrant scarlet autumn foliage.

Melaleuca lanceolata
Moonah

Mature height and spread: 4–10 m **Spacing**: Plant 5 m from other trees or large shrubs **Plant form/habit**: Irregular **Position**: Full sun **Leaf hold**: Evergreen **Benefits**: Flowers, foliage **Flowering time:** Dec–Feb **Flower colour:** Creamy white **Climate**: Warm temperate **Water requirement**: Drought tolerant

The trunks of these trees are always gnarled and contorted and lean with the wind, with a canopy of fine dark green foliage held above. This tree is found most frequently on the Southern Australian coast but can also be seen inland.

There are not many situations where a moonah is acceptable to my design philosophy, but one desirable application is in a beach garden. I first saw these highly sculptural trees growing at Sorrento and Portsea on Melbourne's Mornington Peninsula. Here a unique style of garden design has evolved incorporating the tree. Most gardens in the area are very simple, with rolling lawns dotted with the odd remnant moonah tree and large beds of lavender or agapanthus often bordering the boundary. I fell in love with this relaxed casual style, and try hard to preserve it or reproduce it but only in this southern coastal zone.

The moonah tolerates salt winds, sandy soils and hot dry summers. Being an Australian native the only rules are to plant this tree in the right location and then let it be. It's a great screening tree and can easily be shaped to allow views over its dense mid-green canopy. It is, however, slow-growing and it takes some time before the tree overcomes its awkward teenage years and develops into a mature tree of beauty. If planting new trees, don't prune away the multiple stems. Encourage as much multi-branching as possible – this is what gives the tree its sculptural quality.

Michelia × alba (syn. Magnolia × alba)
White champak

Mature height and spread: 4–10 m **Spacing**: Plant 4 m from other trees or large shrubs; plant 2 m apart for a hedge **Plant form/habit**: Rounded **Position**: Full sun to partial shade **Leaf hold**: Evergreen **Benefits**: Flowers, foliage, fragrance **Flowering time**: Dec–Feb **Flower colour**: White **Climate**: Warm temperate, subtropical, tropical **Water requirement**: Moderate

Michelia × *alba* (now reclassified as *Magnolia* × *alba*) fits into the small tree category, being compact enough to use as an informal hedge or as a single specimen. It has a good large leaf, very typical of subtropical plants. I first saw this tree used as a hedge in Melbourne to great effect: it was dense, a soft pale green colour and its flowers had the most intoxicating scent.
I recommend its use in a small garden or even a courtyard where screening to the boundary is necessary – here the perfume can be concentrated and on a hot summer's night you can be blown away by its aroma. Despite their beautiful perfume, the flowers themselves are fairly insignificant small clusters.

This tree, like most michelias, is sensitive to changes in its environment. These changes could relate to light, soil moisture or temperature. It is not uncommon for this tree to lose a good deal of foliage when it is first planted. This is the tree's way of adapting to its new environment and young leaves are sure to replace the old in a short period of time, as long as the weather is warm enough to promote growth.

This species tolerates a wide range of conditions as long as it has plenty of water – it is happy whether it is humid or dry, and will tolerate the heat and sun. In the southern states I would offer this tree a warm location with protection from wind. The tree can get a bit leggy, so keep it trimmed back by regular tip pruning. It also responds well to a very heavy trimming if it gets out of hand.

Michelia yunnanensis (syn. Magnolia laevifolia)
Michelia yunnanensis

Mature height: 2–4 m **Mature spread**: 1–2 m **Spacing**: Plant 2 m from other trees or large shrubs **Plant form/habit**: Pyramidal **Position**: Full sun to full shade **Leaf hold**: Evergreen **Benefits**: Flowers, foliage, fragrance **Flowering time**: Aug–Sep **Flower colour**: Creamy white **Climate**: Cool temperate, warm temperate, subtropical, tropical **Water requirement**: Moderate

This michelia, now more correctly known as a magnolia, is a compact species with attractive deep green foliage. It flowers abundantly in spring with small creamy white flowers. This plant is more like a large shrub than a tree, and should be utilised as a shrub.

I like it as a dark green background plant to deep flower beds, allowing it to show off its spring flowers and then simply providing background foliage throughout the summer months.

It is tolerant of most conditions but loves moist, rich soil. I find it holds its foliage the best out of all the michelias.

New growth and buds are protected by a soft rusty-brown velvet-like covering of hairs. 'Scented Pearl', a popular cultivar of the species, has more lemon-coloured flowers and its bushes are more compact.

Nyssa sylvatica
Tupelo

Mature height: 10–20 m **Mature spread**: 4–10 m **Spacing**: Plant 8–10 m from other trees or large shrubs **Plant form/habit**: Pyramidal **Position**: Full sun **Leaf hold**: Deciduous **Benefit**: Foliage with autumn leaf colour **Climate**: Cold, cool temperate, warm temperate **Water requirement**: Moisture-loving

For the richest and most spectacular autumn show, *Nyssa sylvatica* cannot be outdone. It turns the most vibrant red that the garden will experience. It needn't be treated as a large tree and grows into a neat conical shape, making it great for smaller gardens. I often use it en masse, planting it in groups or copses of five or more specimens. At my country garden I have planted an entire hill with this tree and it makes for the best autumn show in the garden.

Nyssa will tolerate a hot summer but it is very fussy about moisture; it needs water. It loves a freezing cold winter, which makes it suitable for the cooler areas of Australia, including rural New South Wales, and mountain regions. It needs little pruning – allow it to grow into its natural shape.

Olea europaea
European olive

Mature height and spread: 4–10 m **Spacing**: Plant 8 m from other trees and large shrubs **Plant form/habit**: Rounded **Position**: Full sun **Leaf hold**: Evergreen **Benefits**: Foliage, fruit **Climate**: Cool temperate, warm temperate **Water requirement**: Drought tolerant

Often fruit-bearing trees that have been used for centuries to provide food suddenly enter the world of ornamental plantings. The olive is one such tree; its strong link to the Mediterranean, and in particular the Tuscan, landscape mean it is a favourite when creating gardens inspired by this part of Italy or those that are designed around grey foliage and drought hardiness. It is unfortunate that we in Australia do not often get to see the extraordinarily appealing gnarled, aged trunk of an olive tree – a living sculpture in itself. But as we plant more, over time, future generations will be able to enjoy the history and beauty of this ancient tree.

There are fruiting and non-fruiting cultivars, and some fruiting cultivars do not bear fruit until about 12 years of age. Olive trees prefer soils with a neutral to high pH, so they respond well to applications of lime. These trees also demand a soil that is well drained to thrive. They require little water and love dry hot summers, but will tolerate some humidity and endure a cold winter, including mild frost and snow.

The olive is a highly adaptable tree, responding well to pruning. It can be shaped into a hedge or a topiary, or simply used as a feature tree in the garden. I always prune them and learnt to do so while visiting Italy, where just outside Florence, in Fiesole, I once stayed in a 400-year-old olive grove. The trees were placed exactly 8 metres apart, which seems to be the ideal, and I watched as the old farmer carried out his annual prune. They should be kept open in the middle and pruned back often to five, six, or seven main branches radiating out from the main trunk, at approximately 1.2 metres off the ground.

Pistacia chinensis
Chinese pistachio

Mature height: 8 m (20 years) **Mature spread**: 5 m **Spacing**: 5m **Plant form/habit**: Umbrella **Position**: Full sun, part shade **Leaf hold**: Deciduous **Benefit**: Autumn colour **Flowering time**: Sep–Nov (spring) **Flower colour**: White **Climate**: Cool temperate, warm temperate **Water requirement**: Drought tolerant

One of the toughest of the small deciduous trees, *Pistacia* tolerates extreme heat and drought. It grows to a medium size and is a great shade tree for small to medium-sized gardens.

I am often looking for trees that are not too large. As our garden spaces get smaller, so too does our need for large trees. It's not easy to find trees that will fit into courtyards or small gardens and this tree works well in both of these situations. The spectacular autumn show of orange to red foliage is one of its greatest features.

Phoenix canariensis
Canary Island date palm

Mature height: 10–20 m **Mature spread**: 4–10 m **Spacing**: Plant 10 m from other trees or large shrubs **Plant form/habit**: Palm tree **Position**: Full sun **Leaf hold**: Evergreen **Benefit**: Foliage **Climate**: Cool temperate, warm temperate, subtropical **Water requirement**: Drought tolerant

Although they are technically not trees, *Phoenix* palms are used as a tree in the garden as they give instant height and establish immediately, being able to be transplanted at any height up to 8 metres. I use them mostly at 5 to 6 metres clear trunk height. They are slow to grow and expensive to transport, and you often need to pay for a crane to install them, so they are an investment, but their presence provides instant atmosphere. They are a versatile plant in terms of style, lending themselves to a Mediterranean-style garden as well as a rural Australian one.

Many of the established palms we buy come from old farms in South Australia or Victoria. They are incredibly hardy and require little water once established. Their only enemy is waterlogging – if the roots sit in poorly drained soil they suffer very quickly. They will tolerate salt wind well and I often use them in oceanfront gardens. I prefer to shave the trunks with a chainsaw to create a perfectly smooth trunk, giving a more polished look suitable for city gardens. Male palms are best for courtyards, not leaving the fruity mess below that a female palm does.

Phoenix canariensis will grow in all states of Australia, loving dry as well as humid environments. As a rule, when buying them as an established tree, they come with a squarish root ball approximately 2 × 2 metres and 1.2 metres high. It is very difficult to grow plants other than groundcovers on top of this rootball. Either allow for some blank space, or consider a groundcover such as star jasmine (see page 200).

Pinus pinea
Stone pine

Mature height and spread: 10–20 m **Spacing**: Plant 15–20 m from other trees or large shrubs **Plant form/habit**: Spreading **Position**: Full sun **Leaf hold**: Evergreen **Benefit**: Foliage **Climate**: Cool temperate, warm temperate **Water requirement**: Drought tolerant

I have a wonderful memory of seeing Italian hilltops crowned with these umbrella-shaped pine trees growing out of poor soil, and with very little water. As they thrive in areas like these in Italy, they make a very suitable tree for the drier rural regions of Australia.

With its distinct shape this tree can be recognised from some distance. The nuts it produces are edible – these are the pine nuts we often use in cooking and salads. Its main drawback is its slow growth habit and the fact that it's hard to buy as an established tree. I often use it as a paddock tree in a country garden, as you can simply plant it out and give it very little attention. It also makes for a good driveway tree in the country, again requiring little, if any, water.

Being a dry-area plant, the stone pine doesn't relish high humidity. It has a wide-reaching lateral canopy in the shape of an open umbrella, making it a good shade tree. The bark of this tree is very attractive, with mottled shades of blacks and greys. It likes full sun and even tolerates salt winds.

Platanus × acerifolia
London plane

Mature height: 20+ m **Mature spread**: 10–20 m **Spacing**: Plant 3 m apart for a hedge, 10 m apart for an avenue, or 20 m from other trees or large shrubs for a specimen tree **Plant form/habit**: Spreading **Position**: Full sun **Leaf hold**: Deciduous **Benefit**: Foliage **Climate**: Cold, cool temperate, warm temperate **Water requirement**: Moderate

The fast-growing plane tree is very popular in Australian gardens. It has attractive smooth bark which flakes off into dappled patches of cream, brown and green. Pollarded trees are very sculptural and have an interest of their own in the winter time. I also use *Platanus × acerifolia* as shade trees and specimen trees, for pleached hedges and recently have even shaped them into a living umbrella! This I first saw at the Chelsea Flower Show some five years ago. The tree is manipulated into forming horizontal growth, in this case an umbrella, on top of a trunk that has been cleared to a height of 2 metres. This limits the size of the tree to about 3 metres in height and approximately 3.5 metres in width.

I have also planted plane trees near outdoor tables, making a shady spot to sit and eat. I established a pleached hedge of plane trees at St Ambrose around a swimming pool, locating them 2 metres apart, creating a small hedge of only 3 metres in height. This band of greenery running around the pool on three sides was the perfect area for sitting under when shade was required.

Plane trees are tough but do require some summer water to become established. Even the older mature trees have suffered during the drought, so make sure you have adequate water. They are one of the fastest growing deciduous trees, and can form a suitable scaled drive or avenue within 10 to 15 years. They are versatile in their climatic range, growing well from Brisbane all the way around to Perth. They tolerate the extreme cold and heat, providing they have some water. They offer little autumn colouring, but their leaves and trunks make for a good show throughout the balance of the year.

Pyrus calleryana
Callery pear

Mature height and spread: 4–10 m **Spacing**: Plant 5 m from other trees or large shrubs; plant 2.5–3 m apart for a hedge **Plant form/habit**: Rounded **Position**: Full sun **Leaf hold**: Deciduous **Benefits**: Flowers, foliage with autumn leaf colour **Flowering time**: Sep–Oct **Flower colour**: White **Climate**: Cold, cool temperate, warm temperate **Water requirement**: Drought tolerant

The *Pyrus* genus has proven to be one of the favourite trees of the past 25 years. I first saw them in Dallas, Texas, and thought they would be highly suited to Australia. I researched them when I returned and found an obscure grower, and so began my love affair with the ornamental pear.

In those days the only one I could find was *P. ussuriensis* (see pages 332–33), and I first put it to use as a pleached hedge; in fact, the opening page to my first book featured this hedge. This tree has come a long way since then, with many new species and varieties becoming available. Callery pears are the genus from which most of the new varieties have appeared; 'Red Spire' and 'Bradford' are just two; and 'Chanticleer', which is narrower in form, reaches approximately 5 metres in width if allowed, and holds a tight, narrow egg shape.

The benefit of the callery pear is its upright, tighter branching form. It does not share the problem of branch splitting with its cousin the Manchurian pear. If you desire a pleached hedge the callery pear is not the best type to plant as this type of hedge relies on lateral instead of upright growth. It is perfect for planting on a boundary as a screen or an informal hedge (see page 64 in the Hedges chapter), allowing the plants to grow naturally together. I also often use it as an accent tree in deep garden beds or as a small focal plant in smaller gardens.

This tree has amazing autumn colour and is known for having one of the shortest deciduous periods. It has mid-green glossy leaves and pretty, small white flowers in very early spring. Relatively pest- and disease-free, this tree is also tolerant of dry spells and extremes in heat.

Pyrus nivalis
Snow pear

Mature height: 9 m **Mature spread:** 6 m **Spacing:** 10 m **Plant form/habit:** Rounded **Position:** Full sun **Leaf hold:** Deciduous **Benefits:** Flowers, foliage **Flowering time:** Mid-spring **Flower colour:** White **Climate:** Cold, cool temperate, warm temperate **Water requirement:** Drought tolerant, however best irrigated during extended heat

I have been using the more common ornamental pear *Pyrus calleryana* (see page 328) extensively as both a specimen and hedging tree; this pear shares many of the same attributes, being medium in size, easy to manage and having a relatively non-aggressive root system. The main differences are its sage-green leaf colour, which allows me to use it as a feature or specimen tree in the garden, and its almost weeping habit. The sage green allows it to merge with Australian landscapes, where greys and softer greens are more the colour palette, and *P. calleryana* can often stand out as being too bold. The weeping nature of the way the leaves are held on the branches makes for a softer tree and allows it to fit into a more casual or informal planting scheme. It flowers profusely with white flowers in the spring, but offers little dramatic autumnal colour. Possums love this tree, so be wary where you plant it if you have possum problems. Unlike *P. calleryana* it can have a tendency to lose branches in high winds. It doesn't mind competition from more mature trees and will tolerate reasonably dry conditions.

Pyrus ussuriensis
Manchurian pear

Mature height and spread: 10–20 m **Spacing**: Plant 8–10 m from other trees or large shrubs; plant 2.5–3 m apart for a hedge **Plant form/habit**: Rounded **Position**: Full sun **Leaf hold**: Deciduous **Benefits**: Flowers, foliage with red autumn leaf colour **Flowering time**: Sep–Oct **Climate**: Cold, cool temperate, warm temperate **Water requirement**: Drought tolerant

The Manchurian pear was the first type of *Pyrus* I began using 25 years ago, and at the time the only one I could locate in this country. This pear is the best flowering out of all the *Pyrus* genus, becoming a mass of white flowers in the early spring. The autumn foliage is a brilliant red, and the colder it is the redder the hue.

It is deciduous for longer than the callery pear (see page 328), and has a lateral growth habit, in contrast to the callery's upright form. This lateral growth can lead to the branches splitting in later life, however, so prune

it to deter splitting or multi-trunking in the branch structure.

This pear is a great choice when selecting a deciduous pleached hedge, as it creates a dense mature hedge in five to six years. It is fast-growing and flowers well when clipped into this form. If you are in a hurry to get your hedge established, you can wire branches of neighbouring trees together.

Unlike others in the *Pyrus* genus, this pear grows into a magnificent specimen tree, spreading its branches into a very graceful form and serving as a useful shade tree in the garden. Give it plenty of space so it can form a full umbrella shape.

Quercus canariensis
Algerian oak

Mature height: 20+ m **Mature spread**: 10–20 m **Spacing**: Plant at least 25 m from other trees or large shrubs **Plant form/habit**: Rounded **Position**: Full sun **Leaf hold**: Deciduous **Benefit**: Foliage **Climate**: Cool temperate, warm temperate **Water requirement**: Drought tolerant

Oaks are my favourite trees. I find them so majestic, and if you are fortunate enough to have an established oak tree on your property it will add an amazing sense of history, more so than any other tree. They are often thought of as a quintessentially English tree, and I guess they are – Elizabeth I was standing under an oak at Hatfield House when she was told she was to become queen.

Just as the oak flourishes in England, it does equally well in Australia. During the extended dry period we experienced in 2003–10 I looked around for the tree that was doing the best without water. I perused the paddocks of farms that surrounded my property, and also in paddocks I saw when visiting clients throughout rural Victoria and New South Wales. Believe it or not, the tree I consistently saw to be coping best was the oak! I witnessed many paddocks in which eucalyptus trees were dying but the oaks were flourishing. Therefore I've planted my own garden and fields with oaks.

For an oak that will particularly suit our dry Australian conditions, you cannot go past the Algerian oak. As its name implies, it comes from a dry climate and is very suitable for our long hot summers. It has attractive dark green foliage which, depending on the plant, can remain on the tree as dead foliage in the winter or even sometimes remain partially evergreen. The autumn foliage is not amazing but the tree in leaf possesses a beautiful soft spreading shape.

It is not uncommon for aphids to attack oaks over the summer, which is not noticeable, but towards the end of the season the honeydew left behind by the critters can encourage blooms of black sooty mould late in the autumn. Eventually, however, most of the leaves fall and a new flush renews the tree in the following spring.

It is vital that oak trees have plenty of space to grow. Many Australians plant their major trees too close together and, when faced with a garden of fully grown overcrowded trees, are often reluctant to remove some to allow the beauty of the remaining ones to be seen.

Quercus ilex
Holm oak

Mature height and spread: 10–20 m **Spacing**: Plant 10 m from other trees or large shrubs; plant 3–4 m apart for a hedge **Plant form/habit**: Rounded **Position**: Full sun **Leaf hold**: Evergreen **Benefit**: Foliage **Climate**: Cold, cool temperate, warm temperate, subtropical **Water requirement**: Drought tolerant

This oak is not a common tree in Australia, making it slightly difficult to find. It is well worth the search, though, as it is both incredibly hardy and evergreen.

I first saw the holm oak when working in the South of France, its native area. Here it grows with very little summer rain, making it ideally suited for Australia. The holm oak lines the drive around Centennial Park in Sydney and I have seen it growing in Florida as well. The Miami garden where I saw it growing had used it as a pleached hedge. I have since used it for the same purpose in Australia, but it is slow-growing, so do not expect a perfect hedge in under 10 years.

It will grow in humid conditions and it is a close relative of the cork oak, *Q. suber*. It is not a large oak tree, has small dark green leaves and develops a very sculptural trunk as it matures.

Quercus palustris
Pin oak

Mature height: 20+ m **Mature spread**: 10–20 m **Spacing**: Plant 15–20 m from other trees or large shrubs; plant 8–10 m apart for a copse **Plant form/habit**: Pyramidal **Position**: Full sun **Leaf hold**: Deciduous **Benefit**: Foliage with autumn leaf colour **Climate**: Cold, cool temperate, warm temperate **Water requirement**: Moderate

Pin oaks, with their deeply cut, textural leaves and tight, dense foliage, are one of the more common oak trees grown in Australia. I find them one of the more delicate oaks, however – they seem to require more water than the others and for this reason their appeal is waning slightly.

They are a unique choice when selecting an oak, as they develop a pyramid shape rather than the spreading habit of most oak trees. I use them in smaller gardens where people still want to have oaks. Intense russet foliage appears in autumn and mid-green glossy leaves in the summer. They are easy to prune and it is possible to keep them relatively narrow by tip-pruning regularly.

I have planted them as a large copse at my farm as they love the cold winters and provide me with a strong windbreak once they have joined together. They are reasonably fast growing for an oak.

Opposite: *Quercus palustris, Hedera helix* 'Pittsburgh', *Buxus sempervirens*

Quercus robur
English oak

Mature height: 20+ m **Mature spread**: 20–30 m **Spacing**: Plant 25–30 m from other trees or large shrubs **Plant form**: Rounded **Position**: Full sun **Leaf hold**: Deciduous **Benefit**: Foliage **Climate**: Cold, cool temperate, warm temperate **Water requirement**: Drought tolerant

One of my favourite trees is the English oak. A long-lived tree that endures beyond the human life span, it should be planted and respected for its longevity and its ability to link the past, present and future. Although native to England, it is very tough and will tolerate our dry Australian summers. Slow growing, it takes some years to develop into the magnificent shape we all associate with an oak. Although deciduous it has very little autumn show: at best a dry, pale yellow tinge. It can hang onto dead leaves during the winter. You can avoid this by trying to select your tree during the winter, where you'll be able to tell if a particular specimen has this tendency and choose another one. The leaf litter is one of the best for composting as it breaks down rapidly and provides high nutritional value to the soil.

Many trees create an environment that is hard to grow other plants underneath, but this is not the case with an English oak. Hellebores (see pages 230–32), clivias (see page 358) and hydrangeas (see pages 98–103) all thrive in this environment.

This is a tree that definitely needs space to be appreciated, so resist planting other trees too close, especially in the early years when the garden often looks bare.

Quercus robur 'Fastigiata'
Upright English oak

Mature height: 10–20 m **Mature spread**: 2–4 m **Spacing**: Plant 8–10 m from other trees or large shrubs **Plant form**: Upright **Position**: Full sun **Leaf hold**: Deciduous **Benefit**: Foliage **Climate**: Cold, cool temperate, warm temperate **Water requirement**: Drought tolerant

Often the situation arises where an oak tree would suit a garden but will simply not fit. This is the perfect situation for an upright English oak. This fastigiated form is very contained. It has a conical shape, almost resembling a deciduous pencil pine. For this reason I often use it as a small avenue tree or in pairs in the garden to mark entry points or to frame a feature.

It appears to be tough, like most oaks, needing little water and tolerating extreme summer heat. Its autumn foliage is not exciting, however, often turning brown and seldom a golden yellow. It does have an interesting bare form in the winter, though, showing off its almost perfect conical shape.

Tilia cordata
Lime, linden

Mature height: 20 m **Mature spread**: 8 m **Spacing**: As per plan **Plant form/habit**: Domed **Position**: Full sun, light shade **Leaf hold**: Deciduous **Benefit**: Feature tree **Flowering time**: Summer **Flower colour**: Yellow **Climate**: Cold, cool temperate, warm temperate **Water requirement**: Watering required until established

My favourite *Tilia* is the *T.* × *europaea* (see page 346). With its larger leaf, it's a more graceful tree but it is extremely hard to source in Australia. *Tilia cordata* is the most common species grown and is easily sourced at most tree nurseries. It has a smaller leaf and is denser in foliage. It still has the amazingly scented flowers in summer that always remind me of a European summer. Bees love this tree when in flower.

I often use *Tilia cordata* as a specimen tree, but I prefer to use it as a pleached hedge, which is its traditional use. If you plan to grow it as this type of hedge, buy small plants with a good central leader so it can be trained to have a bare trunk to approximately 1.5 metres high. Train on a wire frame, allowing lateral growth to be clipped to the wires. Clip the tree to form a dense hedge above the bare trunk height.

Tilia cordata will show butter-coloured foliage in the autumn and lime-green new spring growth. I find it burns in extreme heat, especially in its younger years. It must have good drainage, as poor drainage is the main reason for poor health. It requires good watering during the summer months.

Tilia × europaea
European linden

Mature height and spread: 10–20 m **Spacing**: Plant 10 m from other trees or large shrubs; plant 2 m apart for a hedge **Plant form/habit**: Rounded **Position**: Full sun to partial shade **Leaf hold**: Deciduous **Benefits**: Foliage, fragrance **Flowering time**: Dec–Feb **Flower colour**: Cream **Climate**: Cold, cool temperate **Water requirement**: Moderate

I first fell in love with linden trees while visiting English gardens 35 years ago. They are often left to grow into the large trees you see lining grand English or French driveways, or are tamed and shaped into pleached lime walks to enhance smaller garden spaces.

I love them for their flowers in the summer, which boast the most distinctive perfume. They attract swarms of bees. Their leaf is a perfect green, deep lime in colour, and their size is ideal for either standing alone as a large tree or as a clipped hedge. They are extremely flexible and can therefore be trimmed into either normal or pleached hedges. *Tilia × europaea* is one of my favourites mainly due to the lushness and greater texture of its large leaf.

This is a difficult tree to locate in Australia, however – this may be due to its greater sensitivity to the sun than its tougher cousin, *T. cordata* (see page 344). It can burn in the summer sun and normally I would not persevere, but they are worth persisting with as they overcome these difficulties as they mature. In addition, burning only tends to occur for the first five years. They don't seem to reach the majestic size here that they do in Europe, but they are still a large tree by Australian standards. They really only grow in the cooler states or mountainous areas. I have not seen them growing in humid regions.

If training them as a hedge or pleached hedge, be hard on them – prune them severely from a very early age. A pleached linden hedge is often grown on a frame with each lateral leader being trained along horizontal poles, which are placed at approximately 800 millimetre intervals.

Ulmus parvifolia
Chinese elm

Mature height and spread: 10–20 m **Spacing**: Plant 8 m from other trees or large shrubs **Plant form/habit**: Rounded **Position**: Full sun **Leaf hold**: Deciduous **Benefit**: Foliage **Climate**: Cold, cool temperate, warm temperate **Water requirement**: Drought tolerant

Most elms in Australia have been decimated by the elm leaf beetle and have therefore gone out of favour. The Chinese elm is not prone to the beetle, however, and also does much better in drier conditions. Its form and leaf are vastly different to the English elm: it has smaller, darker green leaves and is lighter in density than the other elms. It has a spreading nature with long arching branches that create a tree as wide as it is high. This elm really does not look attractive until it starts to reach maturity. Often clients have admired a Chinese elm of at least 50 years of age and are disappointed when a juvenile elm arrives.

These trees have a tendency to be semi-deciduous during warmer winters and don't completely lose their leaves unless the minimum winter temperature nears zero degrees. A delicate fern-like branch structure with many fine, neatly arranged stems is revealed when all the leaves have fallen. This tree also exhibits an appealing bark, and with age it develops regular mottled patterns with rusty orange patches revealed as the bark flakes away.

This elm is an incredibly tough tree, often requiring no supplementary watering, and will also do well in windy conditions. It will even endure some salt wind. It seems to thrive in a wide range of conditions, growing equally well in Brisbane and in Hobart. If you do not want such a spreading tree, there is a variety called 'Todd' that is more upright.

As this is a difficult tree to tame into any formal structure, just leave it be and it will develop into a nice rounded crown, even though the branching seems wild and irregular from the start. It is pest- and disease-free.

Textural plants

Previous page: *Strelitzia nicolai*, *Cupressus sempervirens* 'Glauca', *Buxus sempervirens*, *Parthenocissus tricuspidata*
Above: *Clivia miniata* 'Belgian Hybrids'

Often a garden can be so homogenous that it becomes boring. One way to avoid this is to inject something radically different into the garden to create interest. Textural plants are ideal for this purpose. A textural plant is one that relies on its interesting and often robust foliage, rather than its flowers, to attract attention. The foliage might be an interesting shape, form or colour – anything really which makes it stand out when placed in a group of plants.

Textural plants may be used to punctuate a flower border and create a focal point. The late great English gardener Christopher Lloyd was a great exponent of this method. He scattered textural plants such as red flax through the flower borders at his renowned gardens in Great Dixter. Visitors today are still drawn immediately to the textural plants before glancing at the rest of the border.

Textural plants can also be used on their own to create a very contemporary planting scheme. A mass planting of yuccas or agaves can fill a planting box in a modern setting to great effect.

I had never really been a big fan of textural plants until recently, when I started using agave (see page 354), often mixed in with softer plants such as star jasmine (see page 200). As I began to experiment more with texturals I discovered many that suited my designs, including cycads (in particular, sago palm, see page 360), which I now use quite a lot to create accents within the garden. I first saw cycads used like this to great effect in Gianni Versace's garden in Miami, Florida, which I was lucky enough to visit while it was being constructed.

Above: *Strelitzia nicolai*, *Cupressus sempervirens* 'Glauca', *Buxus sempervirens*, *Parthenocissus tricuspidata*

Because I admired the deep textural qualities of these plants, I incorporated them in the design for my first garden, which featured in my earlier book *The Boxed Garden*. The planting scheme for this garden was built around two large pots of cycads in the centre of each bed. The effect is stunning as they immediately become the focal point.

Textural plants seem to go in and out of fashion quite frequently, every decade or so. They are very much in fashion at the moment, but who knows how long that will last? When the fashion changes they become despised quite quickly and are often pulled out with much gusto. In the gardens I work on I seem to pull out more texturals which have gone out of fashion than I do the more old-fashioned trees.

Tips for texturals

1. Use texturals to create an impression of depth in a garden.
2. Textural plants can be used to great effect flanking a main entryway. These are the plants to select if you want to make a statement.
3. Choose them as a short-term plant. They do have a habit of coming into and out of fashion quickly.
4. Try to choose textural plants with fleshy leaves. These create a lush, inviting feeling in your garden.

Agave attenuata
Agave

Mature height and spread: 500–1000 mm **Spacing**: 2 plants per m^2 **Plant form/habit**: Rosette **Position**: Full sun to full shade **Leaf hold**: Evergreen **Benefit**: Foliage **Climate**: Warm temperate, subtropical, tropical **Water requirement**: Drought tolerant

The agave has been hugely popular during the last decade. They are stimulating plants displaying interesting texture, colour and shape. The blue-green foliage can be cleverly blended into a garden with other blues, silvers, dark greens or even burgundy foliaged plants.

The reasons for the agave's popularity no doubt include its extreme hardiness, versatility and drought-resistance. It tolerates sun, shade, a wide range of soil conditions (except for waterlogged soils), and will even do well in high humidity. During the worst heat in the southern states it has been known to burn and it is also very sensitive to damage by bruising. If either of these happens, simply cut off the damaged leaves. Over time little offshoots called pups will emerge from the base of the plant and these can be left to develop into a clump or sliced away and planted elsewhere in the garden. The latter is a good way of renewing the plants when they tire.

I use agave as a focal point in a garden bed or en masse mixed with a groundcover such as star jasmine (see page 166). These two coexist extremely well.

Note that agaves do flower but only after some years and shortly before they perish.

Asparagus densiflorus 'Myersii'
Foxtail fern

Mature height: 500 mm **Mature spread**: 900 mm **Spacing**: 700 mm off-centre **Plant form/habit**: Clumping **Position**: Part shade, full shade **Leaf hold**: Evergreen **Benefit**: Foliage **Flowering time**: Autumn **Flowering colour**: Pink to white **Climate**: Cool temperate **Water requirement**: Moderate

This compact clumping plant prefers semi-tropical to tropical conditions but thrives in the cooler states, in the absence of frost and extreme cold weather. It's very drought hardy and will grow in full sun or semi-shaded areas of the garden. Its bright green foliage coupled with its sculptural clump form makes a perfect accent plant in a garden bed. Plant it in areas where you want the eye to be led. It is the perfect textural plant for use in pots. I often use them in a grouping of pots. It is a weed in many areas, so check with your local authority before planting.

Casuarina glauca 'Cousin It'
Cousin It

Mature height: 150 mm **Mature spread**: 1 m
Spacing: 500 mm off-centre **Plant form/habit**: Cascading
Position: Full sun to part shade **Leaf hold**: Evergreen
Benefit: Foliage **Climate**: Cool temperate, warm temperate, Mediterranean, subtropical **Water requirement**: Low, water until established

This is a very drought-hardy Australian native with a weeping habit that lends itself to falling over the edge of post or retaining walls. Its needles grow into a tight, slightly raised clumping form, which will spread out over the ground and cover any object, such as rocks. I use it in mass planting.

Clivia miniata
Clivia, kaffir lily

Mature height and spread: 500–1000 mm **Spacing**: 5 plants per m^2 **Plant form/habit**: Strap leaf **Position**: Partial sun to full shade **Leaf hold**: Evergreen **Benefits**: Flowers, foliage **Flowering time**: Sep–Nov **Flower colour**: Orange, cream **Climate**: Cool temperate, warm temperate, subtropical, tropical **Water requirement**: Drought tolerant

This lily is an essential plant for the shaded garden. It grows where almost no other plant will survive – in deep shade, where there is competition from other roots and where there is minimal water available. To top it off, few other shade-dwelling plants have attractive glossy dark green foliage coupled with large flower heads.

Clivia have strong, thick leathery strap-shaped leaves arranged in a rounded clump once established. During spring, vivid orange flower heads emerge at the end of fleshy stems and hover above the foliage in a bold statement. Recent breeding and selection has developed a creamy yellow form for a more muted colour display and also deep oranges to reds have been developed for those who prefer extra richness in their colour schemes.

Clivia miniata is disease-free, though watch out for the odd chewing insect. In general it is a very low-maintenance plant and has modest water requirements.

Cycas revoluta
Sago palm

✳

Mature height and spread: 1–2 m **Spacing**: 1 plant per 2 m^2 **Plant form/habit**: Rosette **Position**: Full sun to full shade **Leaf hold**: Evergreen **Benefit**: Foliage **Climate**: Warm temperate, subtropical, tropical **Water requirement**: Moderate

The very popular cycad provides a texture to the garden that is not easily found with any other plant. Boasting striking dark green glossy foliage, their rigid fern-like leaves evoke an almost prehistoric feel. The plant is best used en masse where it will provide incredible drama to a garden bed. I also often feature the cycad in a raised planter box, mainly on rooftops or in courtyard situations. I first used them in my home, featured in *The Boxed Garden*, after discovering them in Miami, where they are used extensively in courtyard situations.

Cycads grow best in New South Wales and Queensland rather than Victoria or South Australia. They seem to appreciate the humidity of the north. They will grow in the southern states but seem to have a limited lifespan and need to be replaced every five to eight years. In the southern states they appreciate more sun, while in the northern states they like more shade so they don't burn in extreme summer heat. They appreciate generous irrigation and feeding when the weather is warm. The cycad does not like frost and will burn badly when exposed.

Doryanthes excelsa
Gymea lily

✳

Mature height and spread: 2–4 m **Spacing**: 1 plant per 2 m^2 **Plant form/habit**: Rosette **Position**: Full sun to full shade **Leaf hold**: Evergreen **Benefits**: Flowers, foliage **Flowering time**: Sep–Feb **Flower colour**: Red **Climate**: Warm temperate, subtropical **Water requirement**: Moderate

The Gymea lily is a native plant prized not only for its amazing deep red flowers at the top of long stems that can reach a monumental 4 metres high, but also for its strap foliage. As a foliage plant I use gymea lilies en masse in the garden as they form a tall tight clump of strap foliage that is appropriate in scale for a large garden. I have also used them as feature plants in large planters or pots.

Gymea lilies grow well in sun or shade, their natural setting being in the understorey of eucalyptus forests around the Central Coast of New South Wales, on sandy soils and sandstone plateaus. They thrive in poor soils, which is a plus for Australian gardens; however, good drainage is important. They will grow well in the southern states, even in parts of Tasmania where it is sheltered.

Echium candicans (syn. Echium fastuosum)
Pride of Madeira

Mature height and spread: 1–2 m **Spacing**: 1 plant per m^2 **Plant form/habit**: Rounded **Position**: Full sun **Leaf hold**: Evergreen **Benefits**: Flowers, foliage **Flowering time**: Aug–Dec **Flower colour**: Blue **Climate**: Cool temperate, warm temperate **Water requirement**: Drought tolerant

Echiums are one of those rare plants that are equally as spectacular in flower as out of flower. Their towering spiky leaves rise on the stem far above all other plants, followed by spikes of blue flowers that seem to last the entire summer. They are equally good when added to the rear of the bed or border, or massed on their own to engulf the entire bed. They are very drought hardy and thrive in full sun and hot dry summers. They must have perfect drainage, so I plant them in a raised bed with plenty of coarse sand added.

There is a lot of variability within this species, so select plants with good green-coloured broad foliage, as these tend to produce flowers of a more intense blue. 'Cobalt Towers' is another variety to look out for. A hybrid between *Echium candicans* and *E. pininniana*, it produces huge blue spires like a giant candelabra and should be grown at least once just to see how exciting this plant can be!

Unfortunately Pride of Madeira tends to have a limited life, often dying after only three to four years. If it becomes too leggy after a couple of years, cut it back by 50 per cent as it should then re-shoot and flower prolifically.

Pride of Madeira tends not to like too much humidity, and grows well in the southern states as well as in New South Wales.

Ficus carica
Fruiting fig

Mature height and spread: 2–4 m **Spacing**: 1 plant per m^2 **Plant form/habit**: Rounded **Position**: Full sun **Leaf hold**: Deciduous **Benefits**: Foliage, fruit **Climate**: Cool temperate, warm temperate **Water requirement**: Drought tolerant

Most people associate *Ficus carica* with backyard fruit trees, but I see it as possessing one of the most interesting leaves that can be found in the garden. They are deeply dissected and each leaf is set apart from the other. The leaves have an amazingly sculptural quality. The fig also boasts a very attractive twisted and knotted branch structure, which is visible in the winter. I use it as a stand-alone plant, maybe espaliered against a wall or as a background plant to a garden bed.

As with most fruit trees the fruiting fig is hardy and seems to grow well in all states. It doesn't like extreme cold, however, but will tolerate some frost. Prune it to create the most interesting shape, allowing the branches to weave as much as possible. It needs full sun to grow well and also likes a well-ventilated situation.

Ligularia reniformis (syn. Farfugium japonicum var. giganteum)

Tractor seat plant, leopard plant

Mature height and spread: 700 mm **Spacing**: 600 mm off-centre **Plant form/habit**: Umbrella **Position**: Part shade to full sun **Leaf hold**: Evergreen **Benefit**: Foliage **Flowering time**: Jan–Feb (mid-summer) **Flowering colour**: Yellow-orange **Climate**: Cold, cool temperate **Water requirement**: Regular

The perfectly rounded, glossy green leaves of this plant make it the ideal textural plant for shady gardens in temperate to cool zones. If the winters are too cold, you can cut it back to the ground to survive the climate and frost, but in more temperate areas it will be fully evergreen. It flowers with yellow brushes of flowers in the summer, but its main attraction is the foliage. Water it well in the drier months and avoid full sun.

Philodendron 'Xanadu'
Xanadu

Mature height and spread: 1.5 m **Spacing**: 1 m off-centre **Plant form/habit**: Shrubby **Position**: Full sun, full shade **Leaf hold**: Evergreen **Benefit**: Foliage **Climate**: Warm temperate, tropical, cool temperate **Water requirement**: Regular

This glossy-leaved, compact plant is suited to both shade and sunny positions in the garden, making it a very valuable textural plant. It will do well in both the cool and semi-tropical states of Australia and requires minimal watering in the drier seasons. I find it suits growing close to the trunks of established trees, in the company of plants such as clivias.

Use it in large or small groupings and less as an individual plant. It's a great plant for balconies and indoor gardens.

Phoenix dactylifera
Date palm

Mature height: 4–10 m **Mature spread**: 1–3 m **Plant form/habit**: Palm tree **Position**: Full sun to partial shade **Leaf hold**: Evergreen **Benefit**: Foliage **Climate**: Cool temperate, warm temperate, subtropical **Water requirement**: Drought tolerant

Phoenix dactylifera is the true date-producing palm. It differs in appearance from the Canary Island date palm with foliage that is blue-grey rather than dark green, and the foliage is not as dense. The trunk of the tree is also slenderer. It is one that can be kept at a relatively small size for some years, making it desirable as a foliage plant. I use it as a feature plant to define the corners of a garden bed or even the corners of a feature such as a pond. For me it provokes a very defined style: that of a Spanish courtyard full of citrus jasmine and various textured plants. More than once I have discovered it as a remnant plant in an original Spanish-mission-style garden from the 1920s and 30s. I love it for this reason, as it obviously boasts longevity, and any plant that allows a garden to survive for decades is perfection!

It grows well in all states and will tolerate full sun as well as partial shade. They are a desert palm and therefore require perfect drainage – any waterlogging and they will suffer.

Strelitzia nicolai
Giant bird of paradise

Mature height and spread: 4–6 m **Spacing**: One plant per 2 m^2 **Plant form/habit**: Upright **Position**: Full sun to full shade **Leaf hold**: Evergreen **Benefits**: Flowers, foliage **Flowering time:** Oct–Jan **Flower colour**: Cream **Climate**: Warm temperate, subtropical, tropical **Water requirement**: Moderate

I love this plant for its large fleshy leaves and for its immense scale – I have seen some very old plants reach over 5 metres in height. It dominates any garden and sends a strong message that this is a garden in the style of California or the south of France. I often mix it with gardenias and bougainvillea, usually in a courtyard situation, and accompanying a house with a Mediterranean appearance. Courtyard planting is ideal because the leaves of this plant are ripped to shreds if it is placed in a position exposed to the wind.

The giant bird of paradise is a very hardy plant that will grow in nearly all states of Australia, but will not tolerate frost or extreme cold. I plant them mostly for their foliage because the blooms of this species are pale and less significant. If you want the intense bright orange/blue flowers commonly associated with *Strelitzia*, you need to plant *S. reginae*, its smaller growing relative.

Strelitzia is extremely easy to grow, tolerating shade or sun and even extreme summer heat. It can be prone to woolly aphid attack in a courtyard situation, though, but this is easily controlled with a wipe-down of white oil.

Plants for pots

Previous page: *Sedum* 'Autumn Joy' (in urns)
Above: *Laurel nobilis*

Throughout history, even gardens with plenty of space for many beds of plants have more often than not incorporated some plants in pots. Why is this, when plants clearly do better with access to larger reserves of nutrients in the ground? The answer is quite simply for dramatic effect.

For me pots add another layer of planting that cannot be achieved with garden bed plants. They are often raised due to the height of the pot or its placement, and this elevation helps to focus attention, to give greater importance to the plant and to its location in the garden.

One type of pot I use again and again in my designs is the classical Anduze urn (see page 376). They were first made in the sixteenth century, in Anduze, southern France, and today the same hand-made pottery is still in production. In the eighteenth century they were fashionable among the French aristocracy and landed gentry, who used them for citrus trees, which needed to be moved in the winter to avoid frost. Travellers to France might have seen Anduze urns in the Palace of Versailles. I love using this urn as a link to centuries of gardening history. These pots are the perfect shape for lemons, limes and oranges and to have a garden with Anduze pots full of citrus certainly conveys the feeling of a French garden.

Of course, where natural soil is not available – on rooftops, in courtyards – pots or planters become vital. In small spaces such as these I like to create groupings of pots. These are usually placed in the corners of the space and frequently consist of one large pot flanked by two smaller pots of matching dimensions. These groupings give the effect of abundance without taking up too much precious space. Into these pots you can place formal plantings such as clipped box (see page 374) or informal plantings such as euphorbias (see page 226).

One of the main drawbacks of growing plants in pots is the limited supply of nutrients and water they offer the plant. For this reason they require more maintenance, demanding time and resources for constant watering and regular re-potting in order to replenish the soil. Regular feeding is also necessary, using well-rotted manures or synthetic fertilisers custom-made for potted plants. One trick I have learnt is to paint the inside of the pot with bitumen paint, as this helps retain moisture in the hot summer months.

The ever-durable and reliable buxus is one of my favourites for planting in pots as it is so dependable and

Above: *Pelargonium × hortorum*

can be shaped into any form you wish. Simply ensure it receives enough water and it will thrive in either full sun or full shade. It's a good idea to use *Buxus sempervirens* (see pages 27 and 374) if you live in the southern states and *Buxus japonica* (see page 24) in the humid north. I prefer to shape buxus into a simple sphere and love placing several different-sized pots with varying-sized spheres as a collection together.

Often clients request flowering plants in their pots. I always find this a challenge as very few flowering plants look good when not in flower. One that does pass the test, however, is the geranium. It looks okay when not in flower but of course is a show-stopper when the blooms come out.

Recently I have started experimenting with euphorbia and sedums in pots. Both of these genera have smaller growing specimens that suit life in a pot; the downside is that some of them need to be cut down in the winter. I overcome this by planting bulbs such as tulips that flower in the winter in the same pots.

The greatest collection of potted plants I have ever seen is the grouping that adorns the entrance to Great Dixter. It's a magical mix of evergreen flowering shrubs, some bulbs and a few annuals. This is mad pot-planting en masse. Christopher Lloyd began the tradition, which is now continued by Fergus Garrett, experimenting with unconventional combinations. So while I suggest some of my tried-and-true plants for pots in the following pages, why not continue the tradition of experimental gardening and try different combinations to see what works in your space?

Tips for plants in pots

1. Dark grey or black pots, or pots made from metal, are not suitable for a sunny position in Australian gardens. They cook the roots of the plant, killing it very quickly.
2. If you are going to combine a number of plants in one pot, remember to group together plants that have similar requirements for light, food and water.
3. Consider carefully the mature size of the plant you are planting into the pot. Small plants will be lost in a large pot. Large plants will dwarf small pots and will not thrive.
4. To make sure that pots have the best drainage possible, elevate pots slightly using chucks or 'pot feet'.

Agave attenuata
Agave

Mature height and spread: 500–1000 mm **Plant form/habit:** Rosette **Position:** Full sun to partial shade **Leaf hold:** Evergreen **Benefit:** Foliage **Climate:** Cool temperate, warm temperate, subtropical **Water requirement:** Drought tolerant

When I moved out of home the first potted plant I chose for my apartment was an agave in a very modern pot. I chose it because I could simply cut a stem of the plant and place it in a large pot indoors without soil and with just a bit of water – that is how hardy this plant is.

Agaves tolerate heat, sun or partial shade as well as humidity. However, they may burn in extreme heat and they are prone to bruising if a leaf is badly knocked, for example by hail. Agaves are extremely tough so they are an excellent choice if you need plants that you can leave for weeks on end with little attention paid to them. I like agaves for their sculptural qualities and believe they suit a very contemporary garden or courtyard.

Buxus sempervirens
English box

Mature height and spread: 1–2 m **Plant form/habit**: Rounded **Position**: Full sun to full shade **Leaf hold**: Evergreen **Benefit**: Foliage **Climate**: Cold, cool temperate, warm temperate **Water requirement**: Drought tolerant

One of my favourite plants for pots is English box. This plant is very hardy, tolerating a wide range of conditions from full shade to full sun. English box is extremely malleable, allowing you to shape it into many different forms. Cones and standards can be effective shapes but by far my favourite form for this plant is a neat, clipped sphere.

English box is moderately slow-growing; however, if you are looking for plants to fill out quickly, then regular, feeding and watering over the spring and summer months will help to achieve this.

If you live in subtropical or tropical parts of the country, you don't have to miss out on using box in your garden. Choose *Buxus japonica* (see page 24), which tolerates humidity and has many of the same characteristics as *B. sempervirens*.

Citrus sp.
Citrus

Mature height and spread: 2 m **Plant form/habit**: Rounded **Position**: Full sun **Leaf hold**: Evergreen **Benefits**: Flowers, foliage, fruit, fragrance **Flowering time:** Sep–Jan **Flower colour:** White **Climate**: Cool temperate, warm temperate **Water requirement**: Moderate

An iconic plant for a large pot is a citrus, evocative of the Mediterranean. Citrus grow well in pots and combine beauty in form with productive values.

I most often use orange trees in pots for their intoxicating perfume in the winter and spring months. Like lemon trees, they flourish in containers, clip well and provide elegant, dense, dark green foliage. And of course they provide beautiful and useful fruit.

Citrus are often grown in pots in areas where the winter is cold, as the pot can be brought in to a covered location for the winter and then returned outside for the warmer months. They do need full sun but will tolerate a small amount of shade.

Make sure before you start that the pot you choose is large enough; citrus hate being in small pots. However, if you are going to need to move the pot during winter, you will want to choose one that isn't overly heavy. I would say a minimum diameter of 700 millimetres will serve you well. Keep in mind also that your choice of pot should allow for easy removal for when it's time to re-pot.

Citrus do not like to dry out. Paint the inside of the pot with bitumen paint before planting to help reduce moisture loss through the pot. During the warmer months citrus will need to be watered at least once a week.

Remember that citrus are heavy feeders, so be sure to feed them frequently. I use well-rotted manures but there are good citrus fertilisers on the market also.

Citrus also need constant pruning. Regular tip-pruning will prevent them from becoming woody and help to promote fruiting, while oil will help keep scale at bay. Re-pot them regularly – about every five years.

Ficus carica
Fruiting fig

Mature height and spread: 2 m **Plant form/habit**: Rounded **Position**: Full sun **Leaf hold**: Deciduous **Benefits**: Foliage, fruit **Climate**: Cool temperate, warm temperate **Water requirement**: Drought tolerant

Figs are not often thought of as plants suitable for pots, but I use them in large containers for their highly sculptural, decorative form. I love them when they are out of leaf, when you can see the twisted form of the branches. When in leaf I appreciate the classical leaf form and dark green colour. And when in fruit – well, there's nothing more delicious. A very versatile plant that will reward you for years to come.

Hibiscus rosa-sinensis
Hibiscus

Mature height and spread: 1–2 m **Plant form/habit:** Irregular **Position:** Full sun to partial shade **Leaf hold:** Evergreen **Benefits**: Flowers, foliage **Flowering time:** Nov–Apr **Flower colour:** Multiple **Climate:** Cool temperate, warm temperate, subtropical, tropical **Water requirement:** Moderate

I first saw pots of hibiscus used ornamentally in the South of France. Red-flowering varieties in large glazed pots created a vibrant, exciting focus for the garden. Hibiscus flower well if they are planted in large pots and pruned regularly. They suit a bright garden where warm colours are used. Place the pot in a prominent position and use as a feature in your design. Hibiscus will grow and flower in nearly all environments, but they do prefer full, hot sun.

Use one plant on its own rather than placing it in a collection of pots. This is a feature plant that deserves all the limelight.

Laurus nobilis
Bay laurel

Mature height and spread: 1–2 m **Plant form/habit:** Upright **Position:** Full sun to full shade **Leaf hold:** Evergreen **Benefits:** Flowers, foliage, culinary herb **Flowering time**: Sep–Nov **Flower colour**: Pale yellow **Climate:** Cold, cool temperate, warm temperate **Water requirement:** Drought tolerant

Bay trees are usually used for hedging or as topiary shapes in the garden, but they also make excellent potted plants. They are incredibly tough, requiring very little watering and pruning. I prefer to trim them to a cone or sphere shape, but the classical shape is a standard with a ball atop a clear trunk – this is the shape that has been used extensively throughout history and which features in the most famous collection of bay laurels to be found at Versailles.

Clip the tree once a year, in August or September, just prior to new growth bursts. Try to give it plenty of sun, although it will endure some shade. Re-pot once every five to six years as the bay will rapidly exhaust the nutrients in the soil.

Pelargonium × hortorum
Zonal geranium

Mature height and spread: 300–500 mm **Plant form/habit:** Spreading **Position:** Full sun to partial shade **Leaf hold:** Evergreen **Benefits:** Flowers, foliage **Flowering time:** Sep–May **Flower colour:** Multiple **Climate:** Cool temperate, warm temperate **Water requirement:** Drought tolerant

When I think of Mediterranean-style potted plants I think of pots of warm-coloured geraniums sitting on window ledges or crowded around doorways in small seaside villages. Geraniums are ideal for growing in pots as they love it hot, sunny and dry. Terracotta pots have been used to house geraniums for centuries and are particularly well suited to this purpose as they draw moisture away from the soil, helping to keep the root systems of the plant well drained.

I love using hot red varieties of geranium as an accent in the garden; they command a lot of attention wherever they are placed. I use white geranium as a cooling colour in hot courtyard gardens, often placing them on the edges of steps.

Geraniums need plenty of sun and low humidity – the drier and warmer the better. Treat geraniums as an annual, replacing them every year.

Rosmarinus officinalis
Rosemary

Mature height: 300–500 mm **Mature spread:** 500–1000 mm **Plant form/habit:** Rounded or trailing (prostrate forms) **Position:** Full sun **Leaf hold:** Evergreen **Benefits:** Flowers, foliage, fragrance, culinary herb **Flowering time:** Jan–May **Flower colour:** Blue **Climate:** Cold, cool temperate, warm temperate **Water requirement:** Drought tolerant

It is more and more popular to grow both vegetables and herbs in pots in the absence of a proper vegetable garden. It's possible to have an elegant and productive garden on an apartment balcony, as long as it receives enough sun during the day. Rosemary is one of the essentials for such a garden – its evergreen perennial habit makes it ideal for this type of cultivation.

I choose to grow rosemary in pots not just for this reason, but also for its aesthetic qualities. It is a great plant if you are after a more casual style of potted planting. If you want to grow rosemary as a culinary herb as well as for its ability to be shaped beautifully for a pot, choose the *R. officinalis* species. But try a prostrate variety such as *R.* 'Blue Lagoon' (see page 162) if you are after a casual plant to spill over and drape down the sides of a pot. A benefit of any of the rosemary varieties is that they suit both very modern pots in a contemporary scheme as well as old pots in a more traditional garden.

Rosemary requires good drainage, so use a very open potting soil and provide plenty of drainage material at the bottom of the pot. It's important also that the pot itself has large drainage holes. Rosemary will not tolerate any humidity but thrives in the full, hot sun.

Succulents

Previous page: *Senecio mandraliscae*
Above: *Crassula* 'Blue Bird'. Succulents mix beautifully with olives and other grey foliage plants.

For most of my career, I have focused on using plants that provide interest and beauty from their flowers and their ability to be sculptured into green forms, mostly hedges and spheres. Perhaps the odd sculptural plant, such as *Agave attenuata* or *Cycas revoluta*, was included, but never did I focus specifically or predominantly on textural or sculptural forms for plant material. I believe it's important to remain true to your style but to always allow space to adapt to new fashions, thus remaining relevant and contemporary. One such fashion is the use of succulents in planting schemes.

Succulents have many advantages: they are drought hardy, some requiring very little to no watering; they are highly sculptural, thus creating drama and interest; and they integrate into many of my existing planting schemes, such as Mediterranean and small courtyard designs.

Personally, I don't want them to dominate the garden but rather to create interesting accent points, though for many other designers and gardeners they constitute the entire planting scheme.

Succulents tolerate dry and wet conditions as long as they have excellently drained soils. They are great for coastal conditions and will flourish in almost all of Australia's capital cities.

As we continue to design for lower water usage, succulents will only become more popular. There are thousands of succulents to choose from. My list is limited, but I find these particular species fit with my planting schemes and design philosophies.

Tips for succulents

1. Most succulents like full sun or only partial shade. Ensure you have the correct light levels for your choice.

2. All succulents like well-drained soils and rarely tolerate wet feet. Ensure the drainage is excellent. If need be, add scoria or gravel to open the soil and improve drainage.

3. Nearly all succulents dislike overwatering. Do not plant them in a garden bed with other plants that need excessive watering.

Agave 'Blue Glow'
Blue glow

Mature height and spread: Rosette 600–900 mm; flower pike to 3 m **Spacing**: 1 metre off-centre **Plant form/habit**: Rosette **Position**: Full sun to partial shade **Leaf hold**: Evergreen **Benefits**: Drought tolerant, focal **Flowering time**: N/A (rare) **Climate**: Cold, cool temperate, warm temperate **Water requirement**: Drought tolerant

There are many agaves that look great mixed with softer planting schemes such as perennials and clouding planting schemes. I use the *Agave* 'Blue Glow' mostly due to its neat, compact habit and its smaller size. It works better in groups of three or more, placed as accents to corners of beds or where a statement is needed. Like all succulents, it is extremely tough and prefers full sun, but will cope with limited shade. Its spiky foliage makes it dangerous for children, so place it away from the front of garden beds.

Crassula 'Blue Bird'
Blue bird

Mature height and spread: 800 mm **Spacing**: 400 mm off-centre **Plant form/habit**: Rounded **Position**: Full sun, part shade **Leaf hold**: Evergreen **Benefits**: Foliage, low maintenance **Flowering time**: Winter **Flower colour**: White **Climate**: Cool temperate, warm temperate **Water requirement**: Drought tolerant

Crassulas have a tidy ball-like foliage form, which creates neat tufts of blue succulent leaf. They are extremely hardy and require very little watering. Use them on balconies where automated irrigation is impossible to install. I have mine on my balcony in Sydney, where they only get watered once a month.

They are great for pots or as underplanting to trees where root competition robs all the moisture from the soil. They also look great under olives, complementing their silver colouring.

Senecio mandraliscae
Blue chalk sticks

Mature height: 300 mm **Mature spread**: 600 mm **Spacing**: 5 plants per m^2, 400 mm off-centre **Plant form/habit**: Spreading **Position**: Full sun to partial shade **Leaf hold**: Evergreen **Benefits**: Foliage, winter grower **Flowering time**: Summer **Flowering colour**: White **Climate**: Cool temperate, warm temperate **Water requirement**: Drought tolerant

This low-spreading groundcover has silver-blue leaves resembling strings of chalk sticks, hence the name. I use them in drifts at the front of garden beds, often contrasting them with silver foliage plants such as *Santolina chamaecyparissus* (see page 136) or *Lavandula dentata* (see page 108). It's also low enough to place in front of any clipped spheres in the garden, softening the interface between any paving or lawn and the spheres.

Senecio chalk sticks will take some light shade and tolerate full sun and heat and require little to no watering.

Yucca rostrata
Blue yucca

✳

Mature height and spread: 3 m **Spacing**: 900 mm off-centre **Plant form/habit**: Rosettes, vertical **Position**: Full sun, part shade **Leaf hold**: Evergreen **Benefit**: Foliage, low maintenance **Climate**: Cold, cool temperate, warm temperate **Water requirement**: Drought tolerant

This is one of the prettiest of all *Yucca* plants, mostly due to its blue foliage and graceful appearance especially when it moves with a breeze. The *Yucca* can grow into a small tree but is extremely slow to do so.

I use them as accent plants among mass planting schemes, where they work well as a group or as a single specimen. Place them adjoining grey or silver foliage for maximum effect. They are a bit spiky, so beware of planting in children's play areas.

Maintaining your garden

Water

It almost goes without saying that water has become one of the biggest problems for us as a country, as our population grows and our arid continent seems to be getting drier and drier over the long term. Gardens are more difficult to plan now than they were when I first started.

If there is anything that the recent drought taught us it is that if you want a substantial garden you simply must have an independent water supply such as from large rainwater tanks. Tanks can be the large concrete kind, buried under lawns or drives, or the simple, slender plastic kind that fit down a narrow side of a terrace. My recommendation is that you calculate how much water you can store (and always go for the biggest tank you can fit) and then work out the type of garden or size of lawn you can sustain. I am convinced that despite the odd wet year the drier conditions will prevail and we must plan for them.

Although the layouts and styles of my gardens might not have changed much over the years I certainly have had to change which plants I choose. Fortunately, there is almost always a drought-resistant plant available that can give you a similar look to one of the thirstier ones. It is just a matter of doing a little extra research and perhaps pushing your local nursery to stock the hardier plants you require. In my garden schemes, where I used to plant azaleas, hydrangeas and camellias, I now more often select *Rhaphiolepis* (see page 120), New Zealand rock lilies (see page 218) and sedums (see page 258).

Fertilisers and compost

I try to use natural local fertilisers such as manures and composts rather than synthetic compounds when looking after my gardens. At my country garden we follow an annual ritual of applying well-rotted chicken manure or a bought pellet form of the same each autumn, along with a good 100 millimetres coverage of compost. This helps not only build up the soil structure with organic material but also adds vital nutrients, which are almost always lacking in Australian soils. The only other supplement we use is a liquid seaweed product that helps with root growth, which we apply every second month.

Potted plants need far more regular feeding because of the limited amount of soil from which the often hungry plants can feed. When planting, add a good layer of manure to the bottom of the pot and mix some into the potting mix. Allow space at the top of the plant so you can apply an annual layer of compost. And for potted plants it's a good idea to apply a seaweed solution more often than for other plants, say every month.

Composting is one of the most rewarding activities any gardener can do. Not only does it reduce your contribution to landfill but it creates the richest of fertilisers for the garden. I love seeing the way my household waste is turned into high-quality fertiliser.

At my country garden we made our own compost bay using timber sleepers to form three sides, leaving the front open so the tractor can easily manoeuvre in to turn the heap. You could make a smaller version of the same, using timber sleepers on three sides and then removable planks of wood at the front to prevent the heap spilling. Alternatively there is any number of effective commercial compost bins available.

It takes so little work to make your own compost and the reward is enormous. Simply toss in all your organic kitchen scraps, some garden lime and some hay, then turn it over as regularly as you can, say once a month. You can tell it is well decomposed by its sweet smell and rich, dark colour. Either dig it into the soil or apply as a mulch layer. One word of warning: make sure the compost is not too hot when you apply it, as it can burn the plant material. Water it in well after applying.

Pruning

Pruning is one of the most important jobs a gardener needs to do to maintain a well-proportioned space with happy, healthy plants. And the gardens I design, which incorporate lots of hedges, do require regular pruning because of their formal nature. There is a real art to

pruning hedges and I have to admit I have employed a specialist pruner from the very start. For me, the skill required in accurately pruning a pleached hedge, say, is on par with that of a dentist or surgeon!

Most plants respond to regular pruning by producing more leaf material, which makes it denser and healthier. If pruning your own hedges, never rely on your eye; always run a string line to ensure the cut is level. Nothing looks worse than a sloping hedge. Keep hedges narrow and don't allow them to grow too wide, as it is often hard to get them back to the correct width later on. Try to make the hedge slightly wider at the base as this allows light to penetrate the bottom of the hedge. Prune regularly; don't leave it too long as soft growth responds better to pruning than old wood; in fact some plants do not regrow from old wood.

All of this applies equally to pruning shrubs or trees as hedges. I always shape my trees from a very early age, tip-pruning them as they grow instead of waiting to give them a heavy prune once semi-mature. If you are pleaching trees, work out your clean trunk height from the start – that is the height from the ground to the underside of the canopy – and prune each tree to this height from a very early age. Trim the sides and top regularly, allowing the top to grow slowly to the desired height.

The only times I am disappointed when revisiting a garden I have created years before is when the pruning has not been carried out properly or at all. One client who had an extensive run of pleached hedges decided to save money and prune his own 4 metre high hedges. He did it by eye, leaning back as he made his way up the hedge. The result was a hedge that felt like it was falling in on you. It took years to rescue this hedge and to get it back to being perpendicular with the ground!

Most plants that are not being trimmed into hedges I prefer to grow into their natural shape with very minimal pruning. If anything it may be as simple as not allowing them to hit a building or to grow over a path. The exceptions are when I shape lavenders, santolina or westringia into small spheres in the same fashion as the French, which prevents them becoming woody and helps with flowering.

Mulch

Given the extreme heat and dry summers we have been experiencing, mulching garden beds and trees has become vital in all Australian gardens. The primary objective of mulching is to minimise water loss through evaporation from the soil, but it can also add nutrients back into the soil.

I prefer to use compost rather than bark as a mulch, as compost helps with building up a rich soil profile and I enjoy its dark colour. Many people prefer a bark product, usually made from pine or eucalyptus. I find these often rob the soil of nitrogen and don't contribute to the wellbeing of the soil, although they do a good job of helping to prevent moisture loss.

Renewal, regrowth and change

Gardens are living communities of plants that are subject to the many cycles of nature. They are never truly complete – a truth which can be both a frustration and a joy. A tree might not reach old age for a hundred years, but smaller plants may do so in a year for annuals or three or four years for perennials. Just because some plants are short-lived is no reason to discount them from your garden. Often the lifespans of short-lived plants can be extended by dividing the clumps or bulbs from the parent plants. And this process of helping the plants in your garden to thrive by dividing them, replanting them, pruning them, fertilising them, removing competing plants if necessary – all this only deepens the pleasure of having a well-designed garden.

Acknowledgements

My sincere thanks go to the owners who allowed us to photograph their gardens:

Chrisy and Steve Angelo, Marilyn and Peter Bartels, Robyn and Tony Beddison, Annie and Andrew Bell, Con Bilionis, Lisa and Andrew Blode, Lorraine and Barry Bloom, Helen and Brian Blythe, Liz and Joe Bongiorno, Simon Burke, Geraldine and Andrew Buxton, Louise and Don Carroll, Freda and George Castan, Yve and Vince Chiodo, Chris and Frank Cicutto, Anne Clark, Tracey and Geoff Cohen, Nicola and Alex Commins, Donna and Paul Dainty, Stephen Davis, Kristine and David Deague, Barbara and Malcolm Dumenil, Jill and John Dunlop, Sara and David Ellis, Andrea Evans, Candice and Adrian Field, Fiona and Andrew Fox, Lisa and Peter Fox, Judy and Mark Gandur, Peter Gibson, Jayne Gregory, Wendy and Richard Healey, Margaret and John Hocking, Karen and Graeme Hunt, Robert Kechenovich, Chrissy and Roger Kimberley, Connie and Craig Kimberley, Leona and Ray Malone, Michaela and Matthew McNee, Emma and Andrew Muir, Denise and John Nabb, Michelle Nugen, Selma Oxbrough, Linda and George Paltoglou, Maria and George Phillips, Helen and David Piggott, Liz Polk, Jeanette and John Poynter, Stuart Rattle, Jacqui and Denis Roche, Rosalie and Tony Roosenburg, Karen and Albert Stafford, Carolyn-Michelle and Irek Stawiczny, Sylvia and Roy Tashi, Helen and Jim Thomas, Donna and Michael Tricario, Michelle and John Valmorbida, Kendra and Nick Van De Merwe, Leonard Vary and Matt Collins, Celeste and Socrates Vasiliades, Heloise and Alex Waislitz, Lorna and Quentin Wallace, Margaret and Barrie Webb.

To Simon Griffiths, as ever, for the exquisite images contained in this book.

To Kent Bangay and Angela Payne for their valuable advice and hard work in helping to put this book together.

For their kind assistance: Phil Gange, Zena Bethell, David Glenn and Marc Patterson.

To Teena Crawford for her regular planting suggestions. To Tim Fawcett for looking after my own plants. To Jack Marshall for all his expert plant identification.

To Julie Gibbs for her undying belief in me. To John Canty for his beautiful artwork. To Tracey Jarrett for this high-quality production. To Jocelyn Hungerford and Saskia Adams for proofreading. And to Jane Morrow for her patience and masterful control of my words.

Paul Bangay

Index

Opposite: *Viburnum tinus*

J

K

L

M

N

O

LANTERN

UK | USA | Canada | Ireland | Australia
India | New Zealand | South Africa | China

Lantern is part of the Penguin Random House group of companies
whose addresses can be found at global.penguinrandomhouse.com.

Penguin Random House Australia

This new updated edition published by Lantern, 2021

Front endpapers: *Quercus ilex*, *Buxus sempervirens*, *Ligustrum vulgare*; half title page: *Malus floribunda*; title page: *Agastache* 'Sweet Lili', *Stachys lanata*, *Pyrus calleryana*; opposite: *Syzygium australe*; rear endpapers: *Achillea millefolium* 'Terracotta', *Macleaya microcarpa*, *Calamagrostis × acutiflora* 'Karl Foerster', *Sedum* 'Autumn Joy', *Helenium* 'Wyndley'.

Cover design by Hannah Schubert © Penguin Group (Australia)
Text design by John Canty © Penguin Group (Australia)
Typeset in 11/14 Adobe Garamond by Post Pre-press Group, Brisbane, Queensland
Printed and bound in China

NATIONAL LIBRARY OF AUSTRALIA
A catalogue record for this book is available from the National Library of Australia

ISBN 978 1 76104 310 9

penguin.com.au